T0200641

Living with Disasters

This book is a critical account of the disconnected nature of governance, conservation and livelihood initiatives in the Indian Sundarbans, an active delta that spreads over 25,500 sq. km across India and Bangladesh and lies in the Bay of Bengal. It draws a holistic picture of the disaster-prone delta in eastern India, which is a UNESCO World Heritage Site and also one of the largest tracts of mangrove forests in the world. The author juxtaposes the vulnerable lives and frequently displaced existence of the islanders against the dominant strategies of conservation and development followed by the state.

The book contends that the dominant portrayal of the region as a natural wilderness is not a natural fact but a constructed image. It traces the ecological transformation of the Sundarbans to the early colonial history of reclamation when people settled in that region and huge mud embankments were subsequently built to turn forested wetlands into paddy fields. Much later, with the promulgation of the colonial Forest Act, the imperative of reclamation lost sheen and that of conservation gained ascendancy. Against this specific history of the region, the book provides an account of state interventions around embankment, disaster and displacement in recent times. It explores the islanders' encounter with frequent embankment erosion and disaster over a period of four decades, from mid-1970s to the cyclone Aila in 2009 and its aftermath.

Amites Mukhopadhyay teaches at the Department of Sociology, Jadavpur University, Kolkata. He specializes in environment, development and livelihood issues in India, coastal wetlands in particular and has published in academic journals and edited volumes. His research interests include landscape, ecology and governance, identity politics and institutional histories of science and knowledge making in South Asia.

Living with Disasters

Communities and Development in the Indian Sundarbans

Amites Mukhopadhyay

CAMBRIDGE
UNIVERSITY PRESS

4843/24, 2nd Floor, Ansari Road, Daryaganj, Delhi - 110002, India

Cambridge University Press is part of the University of Cambridge.

It furthers the University's mission by disseminating knowledge in the pursuit of education, learning and research at the highest international levels of excellence.

www.cambridge.org
Information on this title: www.cambridge.org/9781107107281

First published 2016

Printed in India by Thomson Press India Ltd., New Delhi 110001

A catalogue record for this publication is available from the British Library

ISBN 978-1-107-10728-1 Hardback

To my parents
and
the Sundarbans
islanders for their
warmth, courage
and resilience

Contents

Maps and Illustrations

Maps

(All maps are drawn by the author)

Illustrations

(All photographs are taken by the author)

Tables and Charts

Tables

Charts

Glossary

aal	Earthen mounds that separate agricultural lands from one another.
adivasi	Literally means original inhabitant. The word is used to refer to tribes more specifically Scheduled Tribes.
alaghar	Huts built next to a prawn fishery. The huts are meant for the staff employed by the fishery owner to keep a watch over the fishery at night. They are also meant for storing fishing repertoire.
auto	Motor driven three wheeler meant for ferrying people.
bagda	Tiger prawn.
banchbar upay	Ways and means to survive.
bastu	Homestead.
bauley	Tiger charmer and woodcutter.
beldar	Irrigation Department's staff charged with the maintenance of embankments. Beldars are recruited mostly from the adivasi population of the Sundarbans.
bhatbhati	Mechanised boats that connect islands, carrying people and goods.
bheri	Fishery where seeds are grown.
Bidhobapara	Locality where widows live.
bidi	Tobacco rolled in dry leaves for smoking. Bidi is cheaper compared to cigarettes and hence preferred by the poor. In Bengal, use of Bidi has a cultural connotation, for it is associated with the way of life of the so-called subaltern.

bigha	Local unit of measurement, roughly equal to a third of an acre.
bilan	Marshy land where paddy is grown.
Bonbibi	Goddess of forests protecting those who enter the forests in the Sundarbans.
Bonbibi-r Johuranama	The text read out in honour of Bonbibi during Bonbibi puja. The text looks more like prose but reads like verse. The pages of the book open to the right as in Arabic and read from back to front.
bund	Embankments separating land from water, marking the boundary of the fishery and also marking the boundary of one's land.
Catla mach	Indian carp.
char	Sandbars or land formed as a result of silt deposit at the bed of the river.
chouko	Square holes dug on land bordering rivers for adding earth to embankments.
dafadar	Labour supervisor hired by the irrigation department.
Dakshin Roy	Refers to a greedy man-eating half Brahmin sage and half tiger-demon. According to villagers, Bonbibi was sent by god to protect the foresters against Dakshin Roy.
dalal	Broker or middleman benefitting from deals between parties.
diara	Land formed as a result of silt deposit in the Kosi River in the Gangetic delta where settlement started during colonial rule.
down er lok	Literally meaning people from lower echelons of society. Here it is meant to refer to people who live in the southern most islands lying closer to the forests away from places closer to the city.
ghog	Hidden holes in embankments.
jami	Land where paddy grows.
jele	Fisher.
jongol kore	Literally meaning doing jungle. In the Sundarbans it refers to people who go to forests for honey, fish and crab or wood.

jotedar	Refers to big landholder against whom the Naxalites revolted in their land grabbing movements.
kamot	Sharks.
khalashi	An employee belonging to the lowest rung of bureaucracy meant largely for running errands.
kharif	Crop grown during monsoon.
khoti	Little shacks built adjacent to a fishery or on embankments along the river banks where prawn seed transactions take place during the day.
Ma Bonbibi	Literally meaning mother Bonbibi. Goddess is often addressed as mother, the protector against evil force.
mantra	Hymns chanted in honour of the Gods/Goddesses.
mouley	Honey collector who visits the forests.
pabitra mone	Literally meaning 'in pure heart and soul'.
paikar	Traders who buy tiger prawns from fishery owners and sell them to Kolkata-based business people.
panchayat	Three-tier rural self-government instituted by the left government in West Bengal, functioning at village, block and district levels.
pankaj	A particular type of paddy which is a little salt resistant.
patan	Fish trap used in a fishery.
rabi	Crop grown in winter.
raiyat/rayat	Peasants.
saheb	Earlier the term was used to refer to white colonial masters. Now it is used to address one's superior. It is used to refer to white collar government officials, particularly the ones occupying higher positions in bureaucracy.
sajuni	One who readies boat during forest expedition.
sangathan	Literally meaning organization.
Sardar	Tribals or adivasis are referred to as Sardars. Sardar is also used as a suffix to their first name.
Sardarpara	Refers to the locality where Sardars live.

sarkari baboos	A colonial archetype referring to clerks in offices. Used mostly in a pejorative way to refer to people who are not educated gentle baboos but have become one.
up er lok	Literally meaning people from upper echelons of society. In the eyes of the Sundarbans islanders the term refers to people living closer to the city.

Acronyms

ADB	Asian Development Bank
BSF	Border Security Force
CPI	Communist Party of India
CPI-M	Communist Party of India-Marxist
DPD	Development and Planning Department
EIC	East India Company
FB	Forward Bloc
IDPs	Internally Displaced Peoples
IFAD	International Fund for Agricultural Development
IWW	Irrigation and Waterways
MLA	Member of Legislative Assembly
NABARD	National Agricultural Bank for Rural Development
NGO	Non-government Organization
PHE	Public Health and Engineering
PWD	Public Works Department
RIDF	Rural Infrastructure Development Fund
RRI	River Research Institute
RSP	Revolutionary Socialist Party
SAD	Sundarban Affairs Department
SC	Scheduled Caste

SDB	Sundarban Development Board
SDO	Sub-Divisional Officer
SO	Section Officer
ST	Scheduled Tribe
SUCI	Socialist Unity Centre of India
TMC	Trinamul Congress
UK	United Kingdom
UNDP	United Nations Development Programme
USA	United States of America
WBSPB	West Bengal State Planning Board
WWF	World Wildlife Fund

Acknowledgements

In writing an acknowledgement, one wonders if one would be able to do justice to all those involved in making the book possible. It is indeed difficult to name all those who contributed to the publication of the book, which is based on my thesis completed at the University of London and subsequent research in the Sundarbans. I owe my deepest sense of gratitude to the people of the Sundarbans who have shown great resilience, courage and endurance even in times of crisis. I am particularly struck by the generosity and warmth with which they welcomed me into their lives. They reposed confidence in me and volunteered information despite being unsure if my research could change their life for the better. I am grateful to all those who went out of their way to help me, often at no little risk to themselves. I fondly remember the time I spent with Chandan Mondal, Sukumar Mondal, Tapan Mondal, Bhagirath Patra, Shankar Mondal, Mukunda Gayen, Bharat Mondal and Niranjan Bera. I am immensely indebted to my friends Jafar Iqbal and Kalidas Naskar from Basanti who did everything to make my research possible. I am grateful to Shahjahan Siraj and Prabhudan Haldar who ungrudgingly allowed me an access to reports published in the local newspapers like *Badweep Barta*, *Aranyadoot* and *Aajker Basundhara*. Prabhudan Haldar, a retired schoolteacher, has been an archivist in every sense of the term.

I am thankful to the members of the Tagore Society for Rural Development, an NGO based in Gosaba, which has been working among the people in the Sundarbans since the early 1970s. I came into contact with many workers of the Society, but the person who deserves special mention is Ganesh Misra who always took good care of me and made me feel secure even when I was away from home. People like Kanai Sarkar, Sudam Mondal, Radhakrishna Mondal, Sujit Barman, Thakurpada Mondal and Kanai Mondal provided help and guidance during the course of my research. My research would have been impossible without the active help and encouragement of Tushar Kanjilal, the retired Headmaster of a school in Gosaba and the founder of the Tagore Society. Under his guidance, the Society launched a number of social welfare programmes on these islands where infrastructure was virtually non-existent. I am indebted to Tushar Kanjilal for not only having suffered long interview sessions but also facilitating my research by providing valuable contacts in the West Bengal government departments. I had the good fortune of making the acquaintance of Sasadhar Giri, another schoolteacher whose experience and long association with the Sundarbans enriched my fieldwork.

I had the rare fortune of coming into contact with Amitava Choudhury, a doctor who is passionate about his profession. I had the opportunity to accompany him during his visits to different islands of Basanti and Gosaba to treat people of the Sundarbans. It was through him and his profession that I also came to know the Sundarbans. Amitava always put me in touch with people and helped me with much needed information and literature about the region. I owe him an immense debt of gratitude. My sincere thanks are due to my friends Isita Ray, Biswajit Mitra, Sayantan Bera and Dipankar Banerjee for their support at various stages of research.

I am grateful to three anonymous reviewers for their detailed comments and critical observations on the manuscripts of the book. Their suggestions and criticisms helped me look at my manuscript more meaningfully and revise it. This research project could not have been undertaken without the Commonwealth funding awarded and administered jointly by the Association of Commonwealth Universities and British Council in the United Kingdom and the Ministry of Human Resource Development, Government of India. I am thankful to these institutions for funding my research. I am also thankful to the Royal Anthropological Institute (RAI/Sutasoma Award), Mountbatten Trust, Gilchrist Educational Trust, Professional Classes Aid Council, the Leche Trust and Goldsmiths College for financial support when my formal funding was over. My subsequent research (post-PhD) in the Sundarbans was funded by Mahanirban Calcutta Research Group (www.mcrg.ac.in). I remain indebted to MCRG for funding my research on the pre- and post-Aila Sundarbans.

I am indebted to Professor Pat Caplan, my supervisor, for her insights and encouragement throughout the period of research. I am thankful to her for supervising my thesis with care. I have also benefitted from her subsequent comments on some of the chapters of the book. A special thanks is due to those who helped me with their academic advice at various stages of the research in the United Kingdom. Comments from Dr Michael Twaddle and

Dr Amanda Sives at the Institute of Commonwealth Studies helped me prepare for the early stages of my research. I am grateful to Dr John Hutnyk, Dr Eeva Berglund and Professor Brian Morris for their comments as research advisors at Goldsmiths College. Dr Sophie Day and Dr Cris Shore's valuable insights at the writing-up seminars in the department helped me look at my thesis more critically. Equally valuable were the comments of those who attended my presentation at the Anthropology seminar at Goldsmiths and South Asia Seminar series in the London School of Economics. Dr Akhil Gupta helped me with his suggestions during his participation in a workshop at Goldsmiths. I owe a special thanks to Dr Nici Nelson and Professor Lionel Caplan for their comments on the final draft of the thesis. I had also benefitted from the discussion I had with Dr Sarah Franklin at the Anthropology and Science Conference at Manchester University in 2003.

In Kolkata, I benefited immensely from the discussions I had with Dr Anjan Ghosh, Dr Surajit Mukhopadhayay, Dr Dipankar Sinha, Professor Krityapriyo Ghosh, Professor Abhijit Mitra and Professor Prasanta Ray. My book chapters were also presented for discussion at the workshops and conferences organized by institutions such as Calcutta Research Group, Indian Sociological Society, Nehru Memorial Museum and Library, New Delhi. It is perhaps needless to say that the comments and criticisms that I received at these fora only contributed to improving my work. A substantial portion of Chapter 5, 'Beldars, Embankment and Governance: Question of Aboriginality Revisited' was published in Nehru Memorial Museum and Library's occasional paper series under a different title, 'Haunting Tiger, Hugging Ancestors: Constructions of Adivasi Personhood in the Sundarbans' (*New Series 4 Perspectives in Indian Development*, 2013). Chapter 3, 'Governing the Sundarbans Embankments Today: Between Policies and Practices' partly appeared under the title, 'On the Wrong Side of the Fence: Embankment, People and Social Justice in the Sundarbans' (in an edited volume on *State of Justice in India Vol I* by Sage, New Delhi and Mahanirban Calcutta Research Group, 2009) and partly under the title, 'Cyclone Aila and the Sundarbans: An Enquiry into the Disaster and Politics of Aid and Relief' [published in *Policies and Practices series No.26* by Mahanirban Calcutta Research Group (www.mcrg.ac.in)]. I am indebted to the Director and Deputy Director of Nehru Memorial Museum and Library and the Series Editor of the volume on Social Justice and Director of Mahanirban Calcutta Research Group for their kind permission to include sections of the above articles here in the book.

Useful archives that were frequented in Kolkata were the State Planning Board, West Bengal Legislative Assembly and Secretariat Libraries. I am thankful to the officials of these public records offices for their cooperation. I am particularly indebted to Tarun Paine, the Librarian of the West Bengal District Gazetteer, and Aswin Pahari the Librarian of the West Bengal Legislative Assembly for their help with literature and information when I needed them badly. My thanks goes to the officials of the Sundarban Affairs and Irrigation and Waterways Departments for sparing time for interviews. The library staff at the Development Research Communication and Services Centre, Kolkata were extremely helpful in providing relevant newspaper clippings and useful information. I am also thankful to the officials at the local public records offices in the Sundarbans (such as Block Land and Land Reforms Office, Block Office and Irrigation Office).

I remain thankful to my friend and fellow anthropologist Annu Jalais for reflecting on the key issues emerging from my research and helping me with concrete suggestions at various points of time. Her presence in the Sundarbans around the same time I carried out fieldwork made conducting research a pleasurable experience. Annu Jalais' book *Forest of Tigers* published in 2010 made me look at my manuscript much more critically. While writing the book, I fondly remembered my student days at Jawaharlal Nehru University in Delhi, where I had the privilege of being taught by teachers like Professor Yogendra Singh, Professor R. K. Jain, Professor T. K. Oommen, Professor Dipankar Gupta, Professor Maitrayee Chaudhuri and Professor Avijit Pathak. To these teachers of mine, my debts are beyond measure. I am particularly grateful to Professor Maitryaee Chaudhuri who had always insisted that I should publish my research on the Sundarbans. Without her encouragement, the book would have never seen the light of the day.

I am immensely thankful to Nibedita Bayen who was of tremendous help at the final stage of writing the book. Nibedita's help proved invaluable, for it came at a time when I suffered bereavement and had not yet recovered from personal loss on my home front. I am also thankful to Qudsiya Ahmed, Anwesha Rana and Suvadip Bhattacharjee, the publishing team at Cambridge University Press, India, for having gone through my proposal and manuscripts. I am thankful to them for bearing with me even when I failed to meet the deadlines.

As always, Anindita, has been unfailing in her support and suggestions. I am thankful to her and also to little Ahana who realized that writing cannot happen amid fun and frolic and eagerly awaited the day when writing would get over.

Finally, I dedicate the book to my parents who desperately wanted to see the book, but for them the wait proved far too long.

Note on Transliteration

The language spoken in the Sundarbans is Bengali. Like a majority of the North Indian languages, Bengali is derived from Sanskrit and is written in a version of the Devanagari script. Hindi, derived from Devanagari, tends to transliterate into English the vowel 'a' whose sound in Bengali is, in most cases, closer to 'o'. Thus, in Bengali, the terms such as 'panchayat', 'zamindar' or 'dafadar' are generally pronounced as 'ponchayet', 'jomidar' or 'dofadar'. However, convention suggests that these terms be written both in Hindi and Bengali using 'a' instead of 'o'. Hence, I have also followed the conventional practice of spelling these terms with an 'a'.

However, for other local terms such as mouja, chouko, jotedar, jongol and moule, I have used the 'o'. In the case of words like 'jongol', I have also given its English variant 'jungle' as in 'jongol kore' (doing jungle). Although I have mentioned local terms, I have preferred using local terms and their English versions interchangeably throughout the book, for example 'bheri' – 'fishery' or 'bagda' – 'tiger prawn'. However, this rule has not been followed in the case of terms such as 'dafadar' or 'beldar' because their English meanings are long and elaborate. The English meanings of the above terms have been mentioned when the terms are used for the first time in the chapters. I have written all local terms in lower case except for proper nouns. I have avoided italicising local terms. However, the glossary of the book presents the local terms in italics to primarily distinguish these terms from their meanings. I have used all terms without diacritical marks in the text and have added an English 's' to denote the plural. For example, words such as panchayat, bigha or bheri have been used in the plural (i.e., panchayats, bighas or bheris).

Maps

INDIA

Indian Sundarbans

Map 1: Map of India showing West Bengal and the Indian Sundarbans located at the southernmost part of West Bengal.

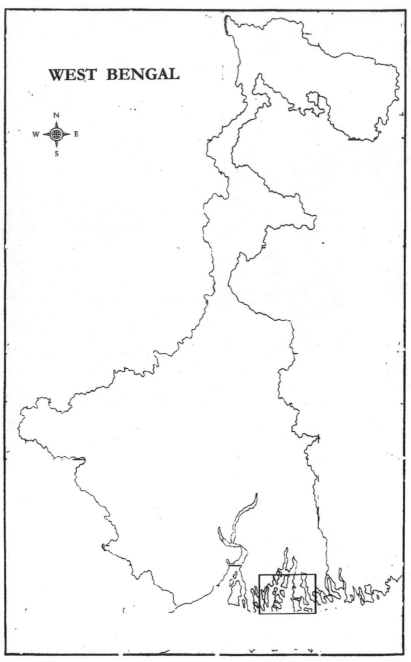

Map 2: The Indian Sundarbans and the area in rectangle denotes the site of fieldwork.

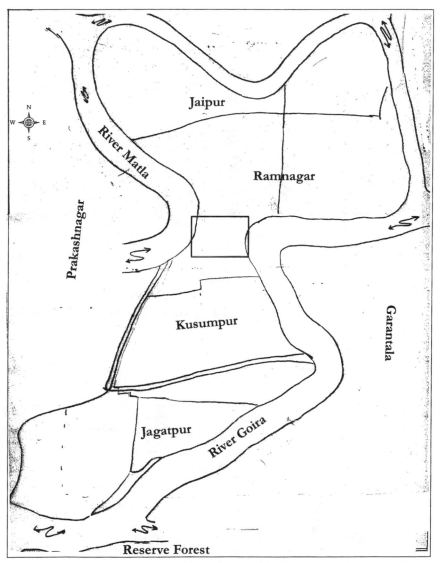

Map 3: Kusumpur island and the area in rectangle shows erosion-prone north
Kusumpur.

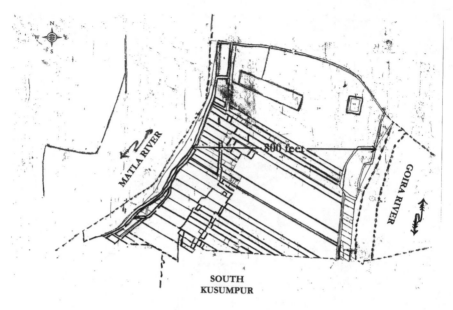

Map 4: Erosion-prone north Kusumpur showing width of the land between the rivers.

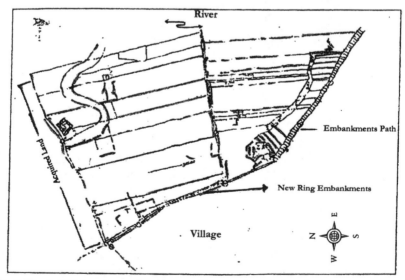

Map 5: The 1,500 feet ring embankment of Garantala and the adjacent 36 acres of land acquired for the purpose of building the new embankment.

Illustrations

1: Erosion-prone embankment

2: Villagers lined up on a collapsed embankment and the village inundated

3: Women catching tiger prawn seeds along the river bank

4: Prawn seeds being counted before the buyers or paikars

5: Prawn fishery with the alaghar being located on the earthen mound of the fishery

6: Fishery with submerged fish trap at the sluice gate

Charts

Minister-in-Charge (Sundarban Affairs Department)
|
Secretary (Sundarban Affairs Department)
|
Special Secretary (Sundarban Affairs Department)
|

Project Director and Member Secretary (Sundarban Development Board)	Joint Secretary (Sundarban Development Board)

Engineering Circle Headed by Superintending Engineer	Planning, Monitoring and Evaluation Division	Social Forestry Division 3 Range officers	Fisheries Division	Accounts Division	Administrative Division 3 Project offices 27 Growth Centres

Division-I 3 Sub-divisions 13 Section Offices	Division-II 3 Sub-divisions 10 Section Offices	Division-III 2 Sub-divisions 8 Section Offices

Chart 1: Organization Structure of the Sundarban Affairs Department

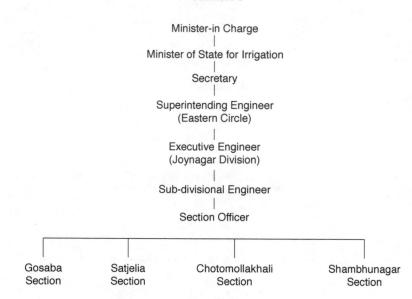

Minister-in Charge
|
Minister of State for Irrigation
|
Secretary
|
Superintending Engineer
(Eastern Circle)
|
Executive Engineer
(Joynagar Division)
|
Sub-divisional Engineer
|
Section Officer

| Gosaba Section | Satjelia Section | Chotomollakhali Section | Shambhunagar Section |

Chart 2: Organization Structure of the Irrigation and Waterways Department

1

Introduction

The Sundarbans embankments: Between land and water

While sitting on an embankment, I started a dialogue with Bhagirath Patra, a resident of north Kusumpur[1], one of the most severely affected parts of Kusumpur island, where the rivers Goira and Matla were eroding banks on the eastern and western sides, respectively (see Maps 3 and 4). Bhagirath, one of the worst victims of land erosion, was now left with less than a bigha[2] of land where he had his house and tiny plot of paddy land. Before I could proceed to interview him, he stopped me and tried to draw my attention to the vast tract of uninhabited landmass of Prakashnagar island right across the river. Pointing towards the opposite bank, Bhagirath told me, 'You know, we used to stay there. We had our bilan jami[3] where the river is flowing now. You can't make out where our house was, it was far beyond the new land that has surfaced and added to the landmass of Prakashnagar. This is the fourth ring embankment[4] that we have seen during our lifetime. Everything is gone. The river and embankment are eating into our strength'.

It would be misleading to consider Bhagirath's narrative as unusual since he shared his agony with many other Sundarbans islanders. The shape and contour of Kusumpur were changing almost every day. Standing on the embankment, it was difficult for an outsider to visualize what the island looked like in the past when the landmass on its western side extended far beyond where the river Matla was now flowing. It was even more difficult to imagine how the new land or char that had emerged out of the siltation process on the other side of the Matla could once have belonged to the people of Kusumpur. But that was how people in Kusumpur perceived their lost land. They believed that the land they lost due to continuous erosion and multiple ring embankments had contributed to the increasing landmass of Prakashnagar. But they could not call this new land their own; they were denied access to it as it belonged to a different panchayat.

As Bhagirath was talking to me, I saw Adhar Mondal, another resident of north Kusumpur walking towards us. He brought to our notice fresh breaches that had appeared in the embankment there. There were now two anxious faces. 'This part' – a huge chunk, almost twenty meters long and leaning towards the river – 'is definitely going to go tonight', said Chandan Mondal who lived in the same area and worked as a labourer at the embankment site. 'How can you be so sure?' I asked him. Chandan smiled and answered, 'I am thirty-three now, we have been living with these breaches since our boyhood'. 'You come to my house tonight', suggested Tapan Mondal whose house was quite close to where the breaches occurred. 'Nothing is going to happen during high tide as the water pressure will be too high now. Tonight low tide will be quite late. If you can stay awake we will come back to see how the river takes away this chunk in the course of its retreat during the ebb tide'. 'At this rate', continued Tapan, 'soon we will have the fifth ring embankment. But I am not sure if we can afford any more ring embankments. The irrigation people told us that the maximum land available between the rivers on both sides is 800 feet'. 'We are about 150 families living in this narrow stretch of land', added Chandan. 'We do not know what fate has in store for us. We are fast losing the land beneath our feet'.

While the people of Kusumpur and the Sundarbans continue to lose the ground beneath their feet and are destined to see their paddy fields flooded with saline water or meet with the failure of their winter crops for lack of freshwater, Sundarbans development remains a widely debated issue, both in the public sphere and in government departments. Today such debates over the Sundarbans are intensified by the acknowledgement that this area is a World Heritage Site, primarily because of the presence of endangered species such as the Royal Bengal Tigers *(Tigris regalis)* and because it is the largest mangrove swamp in the world. The question that follows is: what is the Sundarbans, where is it located?

Sundarbans: The land where nothing settles

The region known as the Sundarbans[5] forms the southern part of the Gangetic delta between the rivers Hooghly in the west of West Bengal and Meghna in the east, now in Bangladesh. Sundarbans, the world's largest mangrove delta, is located between 21° 32′–22° 40′ north and 88° 5′–89° 00′ south. The swamps of the Sundarbans support one of the biggest tracts of estuarine forest in the world. Ghosh (2004) describes the Sundarbans in the following words:

> ... between the sea and the plains of Bengal, lies an immense archipelago of islands. But that is what it is: an archipelago, stretching for almost three hundred kilometres, from the Hooghly River in West Bengal to the shores of the Meghna in Bangladesh ... these islands; some are immense and some no larger than sandbars; some have lasted through the recorded history while others are washed into being just a year or two ago ... The rivers' channels are spread across the land like a fine-mesh net, creating a terrain where the boundaries between land and water are always mutating, always unpredictable ... When these channels meet, it is often in clusters of four, five or even six: at these confluences, the water stretches to the far edges of the landscape and the forest dwindles into a distant rumour of land.
> ... There are no borders here to divide fresh water from salt, river from sea. The tides reach as far as three hundred kilometres inland and every day thousands of acres of forest disappear under water only to re-emerge hours later (Ghosh 2004, 6–7).

The above words capture the Sundarbans in its essence. However, can anything remain as essence in a land where nothing seems permanent? Here we encounter a region being shaped by tidal waves where everything looks transient and fleeting. The forested land which disappears under water every six hours during high tide is the abode of wildlife, while the islands which have erected mud embankments around themselves to prevent their submergence are where people live. Yet, the forests and the islands do not exist in isolation, as rivers keep connecting and disconnecting as they flow around them. Islands are the rivers' restitution, the offerings through which they return to earth what they have taken from it, but in such a form so as to assert their permanent dominion over their gift (Ibid, 7). The area consists of low, flat alluvial plains and is intersected by tidal rivers or estuaries from north to south and by innumerable tidal creeks from east to west. The derivation of the word 'Sundarban', in the words of Pargiter,

> ... is undecided. Several derivations have been suggested, but only two appear to me to deserve attention. One is *sundari*, "the sundari tree," [*Heritiera fomes*] and *ban*, "forest," the whole meaning "the sundri forests;" and the other *samudra* (through its corrupted and vulgar form *samundar*), "the sea," and *ban*,

"forests," the whole meaning "the forests near the sea" ... The second derivation seems to me the more probable (Pargiter 1934, 1; italics as in original).

Thus, the Sundarbans is a place where land is found at the mercy of the rivers. It is the unpredictability of land and water that adds to the natural beauty of the region, a beauty that is a visual treat for visitors and tourists, but a constant source of anxiety and vulnerability for its settlers. This anxiety is manifested in the villagers' narratives which I have presented at the beginning of this chapter, the narratives with which the book opens. One comes across villagers ventilating their angst over their existence, their survival in a land being taken away from them by the rivers.

The Sundarbans: Administrative and social profile

The Sundarbans encompasses an area of over 25,500 square kilometres, two-thirds of which lie in Bangladesh and one-third in India. The Indian part (see Map 1), with which I am concerned in this book, is in the state of West Bengal[6] and covers an area of 9,630 square kilometres. This huge forested area is composed of mangroves, vast stretches of trees and bushes growing in brackish and saline swamps. The Indian Sundarbans, which lies in West Bengal, is spread over the districts of North and South 24 Parganas.[7] The district of 24 Parganas, of which the Sundarbans is a part, remained a single entity until 1986 when, for administrative reasons, it was divided into North and South 24 Parganas. As a result, out of the nineteen blocks that constitute the Sundarbans, six – Hasnabad, Haroa, Sandeshkhali I, Sandeshkhali II, Minakhan and Hingalganj – came under the jurisdiction of North 24 Parganas and the remaining thirteen blocks – Sagar, Namkhana, Joynagar I, Joynagar II, Mathurapur I, Mathurapur II, Patharpratima, Kakdwip, Canning I, Canning II, Kultali, Basanti and Gosaba – became part of South 24 Parganas. An imaginary line called the Dampier-Hodges line serves as the boundary of the Sundarbans and marks it off from the non-Sundarbans parts of the districts of North and South 24 Parganas. This line runs from the south-western part of what is now South 24 Parganas, goes through parts of North 24 Parganas and finally extends beyond West Bengal into Bangladesh. William Dampier, the Sundarbans Commissioner, and Lieutenant Hodges, the Surveyor for the Sundarbans, defined and surveyed the line of dense forests in 1829–1830. In their venture they were helped by Ensign Prinsep's line of dense forests already surveyed in

1822–1823. In 1832–1833, Dampier formally affirmed Prinsep's line in the 24 Parganas. Prinsep's line was renamed the Dampier-Hodges line, which till today determines the limit of the Sundarbans region.

The Indian part of the Sundarbans has 102 islands, of which 54 are inhabited and protected by 3,500 kilometres of earthen embankments and the rest are reserved for tigers. Of the 9,630 square kilometres area that constitutes the West Bengal Sundarbans, 2,585 square kilometres was declared a Tiger Reserve in 1973. There are three sanctuaries in the forest area – Sajnekhali, Lothian and Halliday islands. In 1989, the Sundarbans was declared a World Heritage Site for the following reasons: (1) the Sundarbans is the largest mangrove delta in the world; (2) it is the only mangrove land with tigers to be found anywhere; (3) the Sundarbans possesses the greatest faunal and floral diversity among mangroves of the world; and (4) it serves as a nursery for ninety per cent of the coastal and aquatic species of the eastern Indian Ocean as well as the Bangladesh–Myanmar coast.

The fact less known is that the Sundarbans today is also an abode of about 4.4 million (more than 44 lakh) people. The Sundarbans islands began to be peopled after 1765 when the East India Company (EIC) acquired the civil administration of Bengal (Jalais 2010, 3). However, settlement in the wetlands accelerated around the middle of the nineteenth century when the colonial state in search of revenue leased out large tracts of lands for their reclamation and agriculture. Labourers were hired from Chotanagpur Plateau, present day Orissa (renamed as Odisha) and Arakan coast in Myanmar to reclaim the mangrove wetlands of the delta. This large-scale migration that happened in the Sundarbans during colonial rule was not the first of its kind. Prior to the eighteenth and nineteenth centuries, these mangrove wetlands were inhabited in semi-permanent ways by fishermen, woodcutters, pirates and cultivators (Ibid).

Today the Sundarbans islanders are mostly migrants from other parts of West Bengal or Bangladesh. The islands lying further south (on the margins of the forest) and closer to the Bangladesh border have migrants mostly from Bangladesh, with immigrants still crossing the border and settling into the Sundarbans. These islands on the southern fringes are part of the active delta, being constantly configured and reconfigured by tidal movements in rivers. The areas further up and nearer to Kolkata[8] are parts of the stable delta. The stable delta, just south of Kolkata, has agglomerate, compact settlements that contrast sharply to semi-nucleated, dispersed settlements of the active delta (Banerjee 1998, 184). Nicholas[9] (1963) uses ecology as an analytical category in understanding social structure in two villages in deltaic West Bengal. He makes a comparison between two villages, one belonging to the active delta in Midnapore district and the other belonging to the moribund delta in Birbhum. Nicholas argues that the two villages had distinct social structures affected by their unique ecological location. According to Nicholas, housing pattern, caste structure and political organization of the active delta village contrast sharply with those of the moribund delta village. Nicholas observes,

> ... active delta villages are ordinarily dispersed, ..., with houses quite distant from one another ... moribund delta villages are usually nucleated and compact ... As a consequence ... active delta villages are smaller and in caste composition, simpler than moribund delta villages. Because of their settlement patterns and flooding, social interactions of all kinds – including inter-caste relations – is much less frequent in the active than in the moribund delta villages (Nicholas 1963, 1195–1196).

Nicholas uses a distinctive method of ecology in exploring social relations understood primarily in caste terms. In the Sundarbans the areas that are part of the stable delta are more elevated. The lands here are well irrigated because of their proximity to canals that are not as saline as those of the southern islands. The islands of the active delta do not have elevated ground level and, therefore, have protective earthen embankments or mud quays to prevent daily saline ingress during high tides. In the stable delta or in areas, which are connected to the mainland of West Bengal, prevalent modes of transport are rickshaws, motor-driven three wheelers (often referred to as autos), buses and trains. However, the areas lying further south and surrounding the forests have mechanized boats (locally called bhatbhati) or non-motorized boats (dinghies) as the dominant mode of transport that connects otherwise isolated islands. Most of these islands have brick-paved roads, which only allow cycle-vans (or van rickshaws i.e. three-wheeled cycles with raised platform to carry goods as well as people) to ply. These roads are few and vans ply as far as roads exist. Beyond roads are mud embankments, which serve as pathways connecting one part of an island to another.

In a more or less similar vein, Jalais makes a distinction between the 'up' and 'down' (2010, 5) islands. According to Jalais, 'up' and 'down' are English words used by the Sundarbans islanders as part of everyday Bengali speech to refer to their unique social and geographical location (Ibid). 'Down' islands correspond more or less to places that are part of the active delta lying to the south of the Sundarbans. Not only do these places have low elevations, but also they are cut off from the mainland and hence poorly connected. Being part of the active delta, they are exposed to the risks of cyclonic storm and tidal inundation. By contrast, the 'up' places are those that are part of the stable delta and stand in proximity to Kolkata. Being connected with the mainland they are certainly settled by people who are economically well-off. People living in the 'down' islands often refer to people from areas closer to Kolkata as up-er lok (literally meaning people from 'up' areas). The very phrase up-er lok suggests that 'up' and 'down' are not merely ecological, but socially relevant categories, for it points to the perceptions that people belonging to 'down' islands have of those from 'up' areas and vice-versa. These 'down' islands are the ones lying in proximity to the forests. In characterizing the 'down' islands Jalais states,

> At high tide, when most of the vast expanses of forests go under water, these inhabited islands come alive through communication with each other as sailing between them becomes possible once again. In contrast, during low tide, the forest re-emerges and many of the inhabited islands become isolated once again as riverbeds are left with insufficient water for boats to ply (Ibid, 6).

These are the 'down' islands that characterize the Sundarbans I have described at the beginning of the chapter. My book focuses on people who live on these islands. The opportunities of livelihood for people living on these so-called down islands are very few. People's life on the southernmost islands revolves around water and forest. The Sundarbans is often referred to as the land of jele, mouley and bauley (fishers, honey collectors and woodcutters, respectively).[10] In the absence of heavy or small industries, forest and river remain two significant domains of livelihood. The activities people undertake are not only physically demanding and challenging, but also involve considerable risk. Islanders entering the forest in search of firewood, wood or honey and fish or crab in narrow creeks are often attacked by tigers. According to the forest department's estimate, about 150 people get killed by tigers or crocodiles every year. Women and children belonging to poor families are often attacked by sharks or crocodiles while drawing nets along the riverbanks in search of tiger prawn seeds (*Penaeus monodon*), the largest Indian marine prawn farmed extensively in the region. Agriculture is a source of livelihood for the islanders, but the brackishness of rivers makes agriculture unsuitable and uncertain. As mentioned at the beginning of the chapter, the

Sundarbans embankments, which protect the down islands, are erosion prone resulting in saline ingress and flooding of village lands. Salinity leaves the rice fields unsuitable for cultivation for many years. Winter cultivation is virtually non-existent for want of freshwater. Despite agriculture being an important source of livelihood, a substantial proportion of the farming population belongs to the category of marginal farmers and agricultural labourers. Poor families especially those having very little or no land tend to rely on rivers for marine resources such as fish, prawn or crab.

The islands lying further south and on the margins of the forest are inhabited predominantly by scheduled caste population. As mentioned earlier, these are the islands inhabited by people who once migrated from Bangladesh. The Sundarbans has also a sizeable proportion of tribal population. According to 2011 Census, 40.35 per cent of the whole population of the Sundarbans belong to scheduled caste and scheduled tribe communities (see Tables 1.1 and 1.2 for block-wise distribution of scheduled caste, scheduled tribe and other population). Among the thirteen Sundarbans blocks of the district of South 24 Parganas, Gosaba is one of the southernmost blocks, part of the active delta and down areas, (others being Basanti, Patharpratima, Kakdwip, Namkhana and Sagar) where I carried out my fieldwork (see Map 2). Gosaba block composed of about ten islands, all surrounding the Sundarbans Tiger Reserve, has about 62 per cent of its population belonging to scheduled castes, 9.46 per cent to scheduled tribes and the rest (27.84 per cent) to other backward castes, the so-called upper caste Hindus and Muslims and Christians (as per 2011 Census).

Table 1.1: Block-wise Distribution of Population of the Sundarbans in the District of North 24 Parganas

Blocks	Population	Scheduled Castes	Scheduled Tribes	Others
Haroa	214,401	50,636	12,728	151,037
Minakhan	199,084	60,578	18,564	119,942
Sandeshkhali I	164,465	50,812	42,674	70,979
Sandeshkhali II	160,976	72,300	37,695	50,981
Hasnabad	203,262	51,295	7,492	144,475
Hingalganj	174,545	115,227	12,743	46,575
Total	**1,116,733**	**400,848**	**131,896**	**583,989**

Source: Compiled from Census of India 2011, Primary Census Abstract.

Table 1.2: Block-wise Distribution of Population of the Sundarbans
in the District of South 24 Parganas

Blocks	Population	Scheduled Castes	Scheduled Tribes	Others
Canning I	304,724	144,906	3,710	156,108
Canning II	252,523	52,859	14,910	184,754
Mathurapur I	195,104	68,636	496	125,972
Joynagar I	263,151	102,645	80	160,426
Joynagar II	252,164	85,587	1,046	165,531
Kultali	229,053	104,193	5,672	119,188
Basanti	336,717	119,631	20,060	197,026
Gosaba	246,598	154,584	23,343	68,671
Mathurapur II	220,839	62,342	4,643	153,854
Kakdwip	281,963	97,944	1,836	182,183
Sagar	212,037	56,261	854	154,922
Namkhana	182,830	47,260	741	134,829
Patharpratima	331,823	76,163	2,640	253,020
Total	**3,309,526**	**1,173,011**	**80,031**	**2,057,483**

Source: Compiled from Census of India 2011, Primary Census Abstract.

Living on these down islands – islands in proximity to forests, away from the mainland, geographically inaccessible, lacking basic amenities such as electricity, drinking water and proper roads – assigns a lack of social recognition to these people. The binary 'up' and 'down' (and correspondingly up er lok and down er lok) demonstrates a sense of condescension towards people in these remote locations. It is as if by living in down areas these people – the majority of who were once migrants from Bangladesh – themselves internalize this sense of condescension towards them. They view people from Kolkata or places closer to Kolkata as having wealth and social status, while they feel that they are hapless settlers who have to negotiate their adverse climate and topography to survive and on an everyday basis settle scores with tigers and crocodiles to eke out a living in the Sundarbans (I will discuss this in detail in Chapter 2). The settlers' self-perception as socially and ecologically vulnerable is further reinforced by a lack of development initiatives in the region. This peculiar geography is often cited as an excuse for the lack of effort on the part of the authorities to improve the material conditions of the islanders through programmes such as strengthening embankments, building roads or installing electricity (Jalais 2010, 7).

Entering the field: Island/s at stake

Gardner, in sharing her fieldwork experiences, cautions that fieldwork – usually in some far-flung location – is anthropology's centrepiece, the ultimate transformative experience through which the students of the discipline must pass if they wish to call themselves anthropologists (Gardner 1999, 49). For a person like me, who was born and has spent most of his life in Kolkata, the Sundarbans should not have appeared a far-flung place since Canning, the nearest port of entry into the Sundarbans, is only about two hours journey from Kolkata. However, in the middle-class Calcuttans' world-view the Sundarbans always remains an enigma, a 'wonderland', where tigers stroll and crocodiles swim. In my teens, I often came across people coming back disappointed from their winter visits to the Sundarbans and lamenting that they were unfortunate not to have had a single glimpse of a tiger. But this is not the only time when the urbanites encounter the Sundarbans. The people in Kolkata get to meet and perhaps hear the sad stories of poverty and sufferings of many Sundarbans islanders who arrive in the city in search of employment as domestics in urban households. We also meet such people at tea stalls or roadside eateries where they do odd jobs or run errands. It is through such interaction with them that we experience the Sundarbans in our everyday life. Their presence constantly reminds us how little we know about the land which we occasionally visit as tourists. Thus, the Sundarbans remains a remote place even when many of its inhabitants live right there in the city.

However, when it came to actually negotiating the terrain, the Sundarbans became an even more far-flung place for me. The sheer size of the islands and the expanse of the rivers made me feel completely out of place. I remember the first time I crossed the river Matla at Canning. It was during low tide that I reached the ferry. The water had receded considerably leaving me with no other option but to wade through knee-deep mud. With two reasonably big bags I found myself struggling in the extremely slippery terrain, while people who had started behind walked past me and reached where the boats were anchored in no time. They kept looking back to catch a glimpse of what I was doing. I felt as if I was a stranger to the place, an alien to the people I intended to study. Although my destination was Kusumpur island, my fieldwork virtually started the moment I crossed the Matla. For me, then, carrying out fieldwork meant not only interviewing people and obtaining information, but also coming to terms with the landscape of the region and emulating what others did when they walked on the slopes of the riverbanks or got on or off the boats.

Fieldwork was carried out on Kusumpur island (see Maps 3 and 4) of Gosaba block. Gosaba, composed of several islands, is one of the southernmost blocks, being part of the so-called down areas of the Sundarbans. As mentioned earlier in this chapter, northern part of Kusumpur island was severely erosion prone with the rivers Matla and Goira eroding embankments on both sides (western and eastern sides of the island). There were about 150 families who lived on this truncated and narrow stretch of land. Thus, Nicholas' portrayal of active delta village houses as ordinarily dispersed, with houses quite distant from one another does not seem to apply here. Continuous land erosion and land acquisition robbed the settlers of their lands. This narrow stretch of north Kusumpur looked extremely crammed with houses uncomfortably close to each other. As an urbanite, who had always harboured the idea of a village as being a wide green field, with houses being generally found in relative isolation, I had not found north Kusumpur an example of an ideal village. This chapter opened with the voices and anxieties of villagers which clearly indicated that they were fighting erosion with their back to the wall. Erosion and embankment impoverished them as majority of them remained confined to this narrow stretch and did not have land elsewhere on the island where they could go and resettle.

I conducted fieldwork in four phases. I first started my fieldwork in 2000 and the first phase continued until the end of 2001. I carried out my second phase of fieldwork from 2004 to 2005. I revisited the field for the third time after the cyclone Aila in 2009. I returned to the field after a year in 2010. After Aila this narrow stretch of north Kusumpur looked deserted as many male members of families left the Sundarbans in search of employment in Andaman and Chennai. Kusumpur island was composed of four mouja or revenue villages: Kusumpur, Jagatpur, Ramnagar and Jaipur. These four villages together constituted the Kusumpur village panchayat,[11] which came under Gosaba Block Panchayat. The settlers of Kusumpur island migrated from Medinipur district of West Bengal and Bangladesh. While migrants from Medinipur settled mostly in Kusumpur, those from Bangladesh settled largely in Jaipur. Ramnagar and Jagatpur had both Bangladeshi and Medinipuri migrants. The population of Kusumpur mostly belonged to the so-called backward caste and tribal communities.

People's lives on the heritage site remain vulnerable to the threat and actual occurrence of cyclone and storms in the Bay of Bengal. Two significant cyclones that struck the Sundarbans in recent times were the cyclone of 1988 and cyclone

Aila of 2009. The enormity of the cyclone Aila was such that the Sundarbans remains Aila struck even after four years of the cyclone. Apart from eventful cyclonic disasters islanders' misery and vulnerability is further heightened by a more uneventful and imperceptible process of land erosion and embankment collapse caused by the rivers frequently changing their course. As mentioned above, the islands, which lie on the southern fringes and are part of the active delta, are constantly being configured and reconfigured by the tidal waves of rivers. People on these islands of the Sundarbans live in a state of perpetual anxiety and vulnerability: what if the embankment collapses? Settlement through clearing of forests took place before the natural process of siltation raised the land sufficiently above the water level. Therefore, high earthen embankments were constructed to protect these settlements against daily inundation during high tides. However, the embankments in the Sundarbans spell disaster for the people as breaches constantly occur. Once an embankment collapses, it leaves villages and vast tracts of agricultural land inundated for months before the irrigation department intervenes. When the water finally recedes, the irrigation department further acquires villagers' agricultural lands, houses and ponds[12] for rebuilding the embankment, thereby rendering a sizeable portion of the population further homeless and landless. Agriculture as a livelihood appears unsustainable even though historically Sundarbans was reclaimed for settled agriculture. Sudden embankment collapse and saline ingress not only destroy fully grown crops in the field, but also make the rice fields unsuitable for any further cultivation in years to come. It is these people and their life around eroding and collapsing embankments of the Indian Sundarbans that form the subject of the book.

Disaster, development and marginality of people: What the book aims to do

The book looks upon embankments, the protective mud walls, as providing an entry into the world of disaster and development in the Sundarbans. Literature on disaster tend to view disasters, whether natural, social or technological, as eventful phenomena in people's life. In his book Guha identifies the Union Carbide's deadly gas leak in Bhopal killing several people as a crucial event in engendering public consciousness about environmental degradation in India in recent times (Guha 1998, 1). Guha's observation, though made in a different context in connection with his reflection on the rise in environmental consciousness in contemporary India, tends to convey a very significant idea about disaster as an event, a fundamental interruption to existing social and political life. Disasters – Hiroshima, earthquake in Peru or Turkey, Bhopal gas leak or flooding in Uttaranchal in India – survive in public memory as major epochal events constituting a departure from everyday life (Oliver-Smith 1977a, 1977b, 1991; D'Souza 1986). Oliver-Smith[13] (1996) has identified three main trends in the anthropological literature on disaster namely, behavioural response, social change and political-economic and environmental approaches. He examines these diverse perspectives producing three distinct views of hazards and disasters: disasters as challenges to the structure and organization of society focusing on behaviour of individuals and groups in various stages of disaster and aftermath; disasters as factors of social and cultural change, particularly the transformations imposed on traditional societies by the advanced industrialized world; and disasters as manifestations of human–environment relations resulting from larger historical and structural processes such as colonialism or under-development. Hewitt in his study of calamity or disaster moves away from what he calls the dominant view that treats everyday life and disaster as opposites because everyday life is believed to be orderly and predictable, while disaster is viewed as sudden and unpredictable (Hewitt 1983, 22). Instead, disasters, according to Hewitt, are seen to depend on ongoing social order, its everyday relation to the habitat and the larger historical circumstances that shape or frustrate these matters (Ibid, 25).

Simpson and Corbridge (2006) look at the post-earthquake reconstruction process in Bhuj of western India. They explore the politics of reconstruction showing how refashioning of landscapes in Kachchh is haunted by conflicting accounts of what it is to be a Kachchhi, Gujarati or even Indian (Ibid, 566). The authors show how the post-earthquake rituals and acts of rehabilitation and commemoration reflected the interests of the dominant political groups, which led to the obliterating of many individual or collective memories of the place and the landscape (Ibid, 581). It is in this sense that the earthquake is a break with the past (Ibid). In a subsequent article Simpson shows how this break is sought to be achieved through recourse to a large-scale industrialization that has remained insensitive to local customs, vernacular architectural techniques and traditional livelihoods (Simpson 2007, 937). It is the Gujarat government's twin policy of urban development in Bhuj and rapid industrialization in Kachchh that has unleashed a regime of transformation in the earthquake-torn region.

I draw on the above literature to look at the disaster-prone Sundarbans. Interestingly, here one encounters disaster more as a frequent, mundane and everyday phenomenon. Rivers taking away land in their course or breaches appearing in the walls of the embankment are the realities people live with. The villagers' narratives centring on the broken embankment, with which the chapter opens, demonstrate their vulnerability because the breaches in the mud wall in constant contact with saline water have the potential to inundate and destroy islands instantly. However, one is struck by the ease with which people express their vulnerability or share their individual or collective loss with others. Being in the Sundarbans and listening to people as they narrate their stories, one feels as though disaster is minutely textured into their everyday life. However, to say this is not to ignore the impact of the eventful cyclone Aila that struck the region in 2009. As already mentioned, in terms of its magnitude the cyclone was the severest. But people witness land loss and breaches in embankments and continue to talk about them in the post-Aila Sundarbans in exactly the same way as they did in pre-Aila. This is because the reconstruction process in the post-Aila Sundarbans holds out the possibility of further land acquisition for the purpose of building new embankments in place of the damaged ones. Contrary to the perspective that views disasters and calamities as disruptions to social and political life, my book views disasters as variant manifestations of pre-existing processes and power relations. In other words, disasters serve as a lens through which the nexus between power and development apathy unfolds in the Sundarbans.

The islanders who had survived the cyclone's fury could now become victims of fresh land acquisition in the Sundarbans. The reconstruction project planned by the irrigation department aims to acquire land for the purpose of rebuilding embankment and providing security to the villagers. Villagers are far too familiar with this process of land acquisition and their familiarity is what makes them apprehensive about the government's intention. The islanders in the Sundarbans are not at the mercy of the river alone. Their vulnerability has always been induced by a sense of apathy on the part of the government departments responsible for protecting embankments and providing services in the event of erosion and flooding in the region. Islanders remain marooned for days until irrigation engineers arrive and after they arrive what follows are unilateral decisions about land acquisition for embankment building and protection. Rarely, if ever, are people compensated against their lost lands and houses. Thus, here we come across people who remain perpetually displaced, homeless even at home (Das 2005, 113). Internal displacement or internally displaced people (IDP) has become a theme of research in South Asia (Banerjee et al. 2005). Das defines IDP as people or groups forced or obliged to leave or flee their homes or habitual residence as a result of development, armed conflict, generalized violence, human rights violation or nature-made or human-made disasters (Das 2005, 115). However, in the Sundarbans, the displaced one comes across is somewhat different. The displaced is never at home and constantly on the retreat, backtracking as he loses land to tidal waves and, more significantly, to land acquisition by the irrigation department.

The physical ferocity of natural disasters is relevant in so far as it forces us to look into the questions of vulnerability and marginality of people who inhabit this rich resource site. Intertwined with vulnerability and marginality is a certain portrayal of the region as a natural wilderness and a World Heritage Site, a site for whose conservation islanders' concerns and livelihoods need to be kept at bay. It is as if frequency and inevitability of storm or embankment collapse point to the futility of human settlement in the delta. Thus, here we are faced with a development agenda that puts a high premium on the conservation of wildlife and forested landscape of the delta and in the same breath looks upon people and their livelihood needs as obstacles to the process of conservation. Here we are concerned with the dominant image of the region – as a heritage site and a conservation park – carefully constructed in colonial and postcolonial India. My book aims to explore and interrogate the making of this image (I discuss this in Chapter 2 of the book). The book does this by looking at the biography of the delta in a historical frame tracing the ecological transformation of the Sundarbans to the early colonist history of reclamation when people settled and huge mud embankments were built to turn the forested wetland into settled agricultural land for the purposes of revenue generation. This was the time when landscape and ecology of the region was substantially transformed and shaped by the modern regime of power and knowledge. The book undertakes this historical journey to trace the twists and turns in the process of knowledge production and governance. This historical journey is important to show not only how certain knowledge produced at a particular juncture gained considerable ground in the past, but also how it continues to control our imagination in the present.

However, the Sundarbans was not the only region shaped and transformed by colonial interventions. Colonial rule and its impact on the geography and landscape of South Asia is a fascinating area of research among historians, anthropologists and environmentalists. The ecological history of British India is of special interest in view of the intimate connection that recent research has established between western imperialism and environmental degradation (Gadgil and Guha 1992, 116). Gadgil and Guha attribute this degradation to the industrial revolution and the transformation it brought in the mode of resource use (Ibid, 114). Forest was the first object of exploitation, and Guha

(1989) in his *The Unquiet Woods* explores the colonial forest policy and its impact on the forest resources in the Kumaun Himalayas. The entire ecology of the region was altered by British commercial interests. According to Guha, the genesis of today's Chipko movement could be traced back to colonial exploitative forest policies in the region. Colonial presence and penetration had destroyed the traditional social fabric in the Kumaun region.

Forest continues to attract the attention of the scholars. Sivaramakrishnan (1999) in his *Modern Forests* examines the career of colonial forestry in eastern India. It was under colonial regime of scientific forestry and forest management plan that the forested landscapes of Bengal, from the Darjeeling hills to the mangrove swamps of the Sundarbans, from the eastern edge of the central Indian plateau to the Chittagong Hill tracts, underwent a rapid transformation (Sivaramakrishnan 1999, 1). Sivaramakrishnan deploys the concept of statemaking in unravelling the nature of scientific forestry. He states, 'Forest management was not only predicated upon requisite scientific knowledge but on techniques of validating or valorizing certain knowledge while discounting others. ... The struggle over what knowledge was designated as expertise, who generated it, how it was certified, where it was located and by whom it was practised also became integral to state making' (Ibid, 6). Damodaran shows how the forest landscape of Chotanagpur, imagined in various ways within colonial environmental discourse as wild, remote and pristine, continues to be a site for engendering colonial and nationalist debates on notions of indigeneity (Damodaran 2005, 116). In Rangarajan and Sivaramakrishnan's more recent work (2012), forests remains a site of competing ecological knowledges which in many ways shape our larger sense of the environmental history of colonial and contemporary India.

Water is another site where we encounter colonial science and irrigation. In tracing the history of British irrigation in the United Provinces, Whitcombe argues that the canal irrigation, despite its self-confidence and robust claims about solving the problems of agriculture and irrigation, did not control the vagaries of nature but compounded them (Whitcombe 1995, 258). D'Souza's book *Drowned and Damned* argues how colonial flood control strategies transformed the Orissa delta in eastern India from a flood-dependent agrarian regime into a flood-vulnerable landscape (D'Souza 2006, 2). It shows how practice of flood control is part of a political agenda, deeply implicated in the economic and political calculations of colonial rule.

Mosse (2003) explores the changing ecology and political and communitarian significance of water in relation to tank irrigation in South

India. When we turn away from Mosse's South India to the Sundarbans, we find that it is the saline water that not only rules, but also controls and makes people subservient to its wishes. It was colonial policy of reclamation that controlled and transformed the hydrology of the Sundarbans. When I juxtapose the Sundarbans embankments against the big dams of India especially the way Baviskar examines the making of the Narmada dam in her *In the Belly of the River* (1997), one realizes that practices of flood control, irrigation or hydroelectricity generation stand demystified as an ideological construct driven by certain political and economic calculations. It is this demystification that helps explain why and how imperialist ambitions converge with nationalist aspirations.

If the protective embankment allows us an entry into the world of disaster, it also enables us to witness specific instances of embankment building and flood control undertaken by the government department. Flooding and displacement – whether resulting from cyclones like the Aila or from more silent and persistent processes of erosion at the bed of the rivers – enable us to witness how embankment is protected or built, flood controlled and relief provided. The governance of embankments at the ground level points to interventions involving land acquisitions and displacement of islanders without compensation or resettlement. The realities of river bank erosion, embankment and flooding lead us on to a world of hierarchies, networks of interests and patron-clientelism of different kinds. Thus, governmentality remains central to the understanding of policies and practices of embankment collapse, flooding, rebuilding of embankment and land acquisition. The book focuses on the working of the state departments charged with the welfare of the people and embankment protection and flood control. However, state should not be taken as a free-standing entity, whether an agent, instrument, organization or structure located apart from and opposed to another entity called society (Sivaramakrishnan 1999, 5). The book attempts an ethnography of the state departments' activities in the Sundarbans. An ethnography of the state remains incomplete without an understanding of the presence of the rural self-government or the panchayat institutions at the district, block and village levels (Zilla Parishad, Panchayat Samity and Gram Panchayat respectively) in West Bengal. Panchayats are institutions where governance and partisan interests coalesce in a unique way. A concrete instance of embankment building or protection offers insight into the ways in which negotiations occur between the irrigation department's officials, panchayat functionaries and people in the Sundarbans.

Political context in West Bengal

Since mid-1970s West Bengal had been ruled by a Left-front[14] government consisting of the Communist Party of India-Marxist (CPI-M) and its main coalition partners, which include the Communist Party of India (CPI), the Revolutionary Socialist Party (RSP) and the Forward Bloc (FB). The CPI-M-led coalition had been in power for thirty-four years until 2011 when the Left-front was thrown out of power by the Trinamul Congress (TMC)[15] under Mamata Banerjee as the Chief Minister of the new government. This long stay in power or what the former Chief Minister of the Left-front government described as an 'instance of unprecedented electoral success' (Basu 2000, 1) can be largely attributed to the way it has mobilized its support base in rural Bengal. Two significant policies that had helped the Left-front government to consolidate its support base were land reforms, aiming at land redistribution for the poor peasants in a democratic manner; and panchayat, an experiment in democratic decentralization. Historically, the communists were engaged in various land struggles in the pre- and post-independence era when they chose a revolutionary path to assert the rights of the landless against the zamindars or jotedars. However, the communist party that came to power in 1977 was more reformist than revolutionary (Kohli 1990, 367), aiming to radicalize the rural landscape through democratic and electoral means. According to Kohli, the emergence of the left as a reformist, but disciplined government was seen as a welcome relief after years of political uncertainty and anarchy which people experienced as a result of the breakdown of political order during the erstwhile Congress rule (Kohli 1997).

For thirty-four years, the Left-front was in power both at the state and three-tier panchayat levels. Over these years, the front had also developed entrenched party structures that ran parallel to the panchayat institutions at three levels. In fact, the left parties had developed elaborate party structures from the national level down to the local level with each layer corresponding roughly to administrative units such as province, district, block and village. These elaborate left-party structures explain the left dominance in West Bengal. Interestingly, the Trinamul Congress had developed equally elaborate party structures to counter left-party machinery at all levels. The result was a landslide victory for the Trinamul Congress in 2011 when the party defeated the front at the state level. The Trinamul Congress had also maintained its dominance in the panchayat institutions in the recently concluded panchayat elections in 2013. Thus, it is the institutions of decentralized governance together with entrenched party structures (of the left to a large extent) that provide the context against which I understand the dynamics of development as it unfolds in the Sundarbans.

Eroding lives: Victimhood and/or agency

The book does not simply restrict itself to a critique and exploration of the dominant image of the Sundarbans and the development paradigm integrally connected to that image; it also turns to the islanders, the lives that erode. However, the book views the islanders not solely as hapless victims of disaster and developmental apathy, but as those imbued with a sense of agency to fight adversities and lead their life. The focus here is on livelihood strategies of islanders in the capacities of embankment workers, beldars (government-appointed people in charge of embankment maintenance), prawn seed collectors and fishers. However, in doing this the book argues against what Scott calls the 'standard narrative' (Scott 2000, viii) whereby the colonizer, the market and the state are the agents of ecological degradation while indigenous people are nature's natural conservators. Baviskar has expressed similar anxiety when she encounters what she calls a 'dissonance between the depiction of tribal people in scholarly writing on the subject, and the everyday lives of the tribal people' (Baviskar 1997, vii). The question she raises is whether the lived reality of tribal people today allows the formulation of a critique of development (Ibid). Paradoxically, the indigenous or the local's lived world is found inescapably shaped by their increasing encounter or engagement with the development machinery no matter how alienating or disempowering it is. Keeping the above perspective in mind, the book reflects on the livelihood practices of the local communities in the Sundarbans. The book does not subscribe to the dominant perspective in South Asian anthropology that seeks to understand communities by pigeonholing them into caste, tribe, religion or other indigenous identities. Instead, the book looks at community dynamics at two levels.

- Communities shaped by the eroding landscape and ecology of the delta. Continuous erosion and shrinking space is what puts the islanders on the back foot. Erosion provides a clue to how people construct their notions of space. The islanders have a unique sense of attachment to the land lost and land gained somewhere else. A crisis, actual or perceived, therefore produces a sense of solidarity amongst the settlers of a space.
- Collectives shaped by their encounter with governmental agencies at the local level. Following Agrawal (1999) and Chatterjee (2004), my book critiques indigeneity as the basis for understanding communitarian

identities. Practices of communities such as embankment workers, beldars, prawn seed collectors and fishers crystallize in the context of governmental activities surrounding embankment maintenance, construction and flood control. In a land where livelihood options are limited and non-existent, the embankment, which is a source of disaster for the islanders, also provides them with opportunities of livelihood. Thus, it is a prospect of livelihood, even in times of crisis and vulnerability, which makes visible transactions and negotiations between workers, beldars and fishers on the one hand and local levels of state machineries on the other. The book documents these negotiations only to demonstrate that in a muddy and slippery terrain community ties tend to become contingent and fleeting.

Here one is reminded of Lahiri-Dutt and Samanta's (2014) recent work on char land of the Damodar river in Bardhaman district of West Bengal. The book is yet another story of people on a land that rises from the bed of the river, a land that floats on water (Lahiri-Dutt and Samanta 2014, 1). The book is a telling commentary on the identity, agency and livelihood strategies of communities as they negotiate their 'hybrid' and transient world.

Influenced by the environmental perspectives on South Asia, the book in many ways carries forward the tradition of critical thinking on landscape, ecology and governance in the subcontinent. The book is an account of people's life as it revolves around eroding river banks and frequent embankment collapse in the Sundarbans. It explores islanders' encounter with embankment and disaster over a period of about four decades (i.e., from mid-1970s to the cyclone Aila in 2009 and its aftermath). In essence, the book stands at the interface of anthropology, history and environment studies. The strength of the book lies in its ability to make embankments, the long stretch of landmass encircling the islands – an indispensable constituent of ecology and landscape of the deltaic Sundarbans – an object of anthropological knowledge. Far from being merely a physical landmass separating land from water, the embankment emerges as a reality that provides access to political exigencies of governance, contested ecological and development visions and, last but not least, islanders' everyday life.

Organization of the book

The book is divided into seven chapters. The first chapter has introduced the Sundarbans as a land of water and tides where nothing settles. The region is also described as a forested landscape, a wildlife sanctuary and a heritage site whose conservation is of crucial importance in contemporary India. The Sundarbans is also a place where people have settled. The chapter has focused on the marginality of the islanders in a disaster-prone and flood vulnerable region and has argued that people's vulnerability is induced not simply by natural disasters but by the scant attention being paid to problems people face. Situating the Sundarbans in the wider environmental discourses in South Asia, the introductory chapter sheds light on what the book aims to do. At the end, it has provided an outline of arguments in the subsequent chapters.

In the second chapter, I argue that the Sundarbans as a forested landscape and an abode of wildlife is not a natural fact but a carefully constructed image. Accordingly, the chapter interrogates the construction of the dominant image of the Sundarbans as a national park to whose conservation people's livelihood needs appear as obstacles. It traces the Sundarbans' journey from being a wasteland to a wonderland. In seeking to explain why people continue to remain marginal in this heritage site and have their existence at stake, the chapter embarks upon a journey looking at the region in retrospect. It revisits the historical moments when the Sundarbans wetlands was being reclaimed by colonial rule for the purposes of settled agriculture and revenue generation, when the landscape and environment began to be shaped by the modern regime of power and knowledge. The chapter traces the twists and turns in the process of knowledge production and governance. It focuses on significant historical moments, both in colonial and post-independent India, when practicalities of state making went hand in hand with the image constructed of the region. By doing this, the chapter unravels conservation as a political project.

The third chapter follows from the second chapter and examines governmental policies towards embankment, erosion, flooding and people's development in the present day Sundarbans. I argue how the primal image of the Sundarbans as a natural abode of wildlife and a transient and perpetually unsettled land, coupled with the imperative of conservation continues to inform the shaping of the policy discourses. In other words, government's disaster management policy is conspicuous by its absence. The chapter also turns attention away from the policy arena to concrete instances of disaster management, embankment building and protection. I intend to show how apathy is manifested in the disaster management or flood control interventions undertaken by governmental agencies working at the ground level. The chapter dwells on the cyclone Aila and the politics of post-Aila relief and aid when political parties scrambled to expand their political base among the Aila-struck population. The irrigation department's post-Aila flood control policy advocacy has largely remained on paper. I show how islanders are left at the mercy of developmental apathy. Building on governmentality, the chapter shows how embankment figures as a governmentalized reality in both statist rhetoric and practices.

The fourth chapter provides an overview of Kusumpur island highlighting in particular the eroded stretch of north Kusumpur village. I throw light on the nature and course of settlement and people's location on the island. The chapter focuses on north Kusumpur where people live under the perpetual threat of displacement. Their vulnerability, actual or perceived, clubs them into a single category (community of vulnerable in this case). The chapter here situates the specific empirical case of north Kusumpur in a wider historical context, delving into the colonial riparian rules and the role it played in producing a vulnerable landscape and settlement elsewhere in the Gangetic delta (the Kosi diara region is discussed in particular). The chapter also argues that in a vulnerable landscape where nothing settles, the bonding people have with each other tends to be fleeting as well. A collapsing or collapsed embankment entails displacement and misery for the islanders, but paradoxically, building that embankment presents them with the possibility of a livelihood. Thus, it is a prospect of livelihood, even in times of crisis and vulnerability, which makes possible transactions and negotiations between embankment construction labourers, beldars and fishers on the one hand and local levels of state machineries on the other. The fourth chapter documents the activities of islanders around embankment repairing and building. The focus here is on islanders' agential role manifested in various acts of survival, subversion and protest. The chapter finally examines people's livelihood strategies in the light of the weapons of the weak perspective.

The fifth chapter focuses on the adivasi population, their identities as forest clearers/tiger chasers and beldars, the irrigation department's staff charged with embankment protection in the Sundarbans. The chapter revisits colonial indentured labour history and the colonial stereotyping of the adivasi as wild people to understand adivasi Sardars' position in the Sundarbans today. The so-called tiger chasers are tiger victims as well. The chapter examines beldars' position in the light of the non-adivasi perceptions of the adivasi sardars. Frequent embankment erosion and collapse is often attributed to the negligence and laxity on the part of the beldars, who are perceived as having cut a sorry figure for their illustrious forefathers hired as indentured labourers for reclaiming and building the Sundarbans. The chapter juxtaposes present day beldars' (as government functionaries) narratives against this romanticized portrayal of adivasis as aboriginals having uncanny physical strength and courage. Illustrating the predicaments of the present day beldars, the fifth chapter examines indigeneity as the basis for building communitarian identities.

In the sixth chapter, I focus on the practice of tiger prawn seed collection and trade, a major source of livelihood in the Sundarbans. The chapter focuses on women prawn seed catchers in the light of their livelihood being portrayed as a threat to biodiversity and contributory to embankment erosion. The government departments have come down heavily on the women fishers catching prawn seeds along the riverbanks. However, in trying to understand the dynamics of prawn seed collection, the chapter lays bare a whole network of interests around prawn collecting, farming and trading. The chapter unravels diverse layers of the phenomenon that I choose to call prawn politics. Interestingly, the state, which views prawn catching as a threat to the unique ecosystem, is found implicated in prawn politics. The chapter focuses on various activities such as fishery making, embankment breaking, cross-border trading and political parties controlling fisheries, which are constitutive of prawn politics. The chapter shows how the Sundarbans embankment remains a pivot around which revolve prawn trade and prawn politics.

The seventh chapter concludes by reflecting on how lives continue to erode despite protective embankments and islanders' vulnerability to hazard persists. A heritage site, known for its tiger and wildlife reserve, becomes a spectacle of marginality, destitution and suffering. The book draws the attention of policymakers and concludes with suggestions for changes in policies towards the Sundarbans development.

Notes

[1] For the purposes of confidentiality Kusumpur is the name I gave to the island where I carried out my fieldwork.

[2] Bigha denotes a local unit of measurement, which is roughly equal to a third of an acre. Although bigha does not constitute a part of the official system of measurement, people prefer to use this for calculating their cultivable land and other immovable property.

[3] Jami or land is mainly of two types: bilan and bastu. Bilan refers to marshy land where paddy is grown, whereas bastu denotes the land where one's house is situated.

[4] Ring embankment refers to a particular way of repairing or reconstructing a stretch of breached or collapsed embankment. Ring embankment is further explained in the third chapter of the book.

[5] I use the word Sundarbans, generally referred to in the plural, to denote the region composed of forests, inhabited mainland and islands and water bodies. However,

the administrative departments such as the Sundarban Affairs or the Sundarban Development Board, discussed in subsequent chapters, refer to the area in the singular.

[6] Since the Indian Sundarbans is situated in the state of West Bengal, prefixes such as Indian Sundarbans and West Bengal Sundarbans will be used in the book interchangeably.

[7] Parganas here refers to revenue areas or villages.

[8] Calcutta has been renamed Kolkata. The book will use Calcutta and Kolkata interchangeably.

[9] Nicholas' work is seminal, because way back in the 1960s he coined what he calls *method of ecology* to study social relations of villages located in two distinct ecological settings. For details see Nicholas (1963).

[10] These are the people who represent the population of the Sundarbans. In popular imagination and also in literary fictions the Sundarbans is often portrayed as the land of jele, mouley and bauley.

[11] The units of self-government introduced were the three-tier panchayat system composed of the Zilla Parishad, Panchayat Samity and Gram Panchayat at the district, block and village levels, respectively. All the three tiers are constituted through elections.

[12] Individual ponds are the only source of freshwater in the Sundarbans.

[13] For a wider discussion on the above classification see Oliver-Smith (1996).

[14] I use the term left with capital 'L' only when I refer to the Left-front government. In all other cases I shall be using left in lower case.

[15] The word Trinamul in Trinamul Congress will be spelt as 'Trinamul' throughout the book, unless otherwise spelt as 'Trinamool' in citations.

2

From Wasteland to Wonderland
The Making of a Heritage Site

Whose heritage site?

The first chapter has already introduced the Sundarbans as a World Heritage Site and described the life of its people as one of perpetual anxiety and uncertainty. What is it that makes people's lives so agonizing and uncertain in a place that is globally famous? The answer partly lies in the question posed. The Sundarbans is the abode of tigers, deer, snakes, crocodiles, dolphins, turtles and innumerable valuable plants and marine resources, whose protection and conservation has attracted considerable global funding in recent years. Such a conservation drive is based on the implicit assumption that the sustainable development of the Sundarbans is possible only if the natural habitat of the region is left to grow without hindrance. In other words, people's settlement and pressure on the unique ecological system is a cause for concern among the conservationists. It is to this representation of the Sundarbans as primarily a place of natural wilderness where humans occupy a position of secondary importance that I turn to in this chapter.

However, this image of the Sundarbans as natural wilderness not quite suitable for human settlement is not a natural fact since time immemorial. The biography of the region has been shaped and reshaped by colonial statemaking (Sivaramakrishnan 1999) involving complex negotiations at various local and trans-local levels. Sivaramakrishnan develops the idea of statemaking and employs it in understanding the emergence and institutionalization of modern forestry in colonial India. I draw on this concept to suggest that colonial statemaking in the Sundarbans proceeded through different phases and each phase produced a particular image of the region. This chapter embarks upon a historical journey revisiting the early period of colonial rule when the Sundarbans emerged as a wasteland over which conflict occurred between the landlords and the colonial state. This chapter goes back to the time when reclamation and human settlement in the Sundarbans was considered an urgent necessity by colonial rule. The chapter also looks at the post-reclamation history of the delta when the formulation of Forest Act and the modalities of gazetteer writing began to portray the region differently. What I attempt here in this chapter is not a chronological history. Instead, I will touch on four significant moments, both in colonial Sundarbans and in post-independent Indian Sundarbans, which played a crucial role in consolidating the image of the region as an abode of wildlife, disaster prone and inhospitable for human settlement. By revisiting the four moments, I will show how this image has gained ground and how a considerable array of literature produced on the Sundarbans in post-independent India tends to treat this dominant portrayal of the region as naturally given without carefully analysing how it was constructed. The chapter starts with a brief account of the reclamation history of the Sundarbans under colonial rule, highlighting the process of statemaking in the region. In the absence of a clear-cut boundary of the Sundarbans forests, there was a conflict between the zamindars who claimed ownership over large tracts of forest and the government claiming the same as state property. Thus, the status of the Sundarbans presented the colonists with a dual problem, namely how to wrest the entire forested area from the control of the zamindars and how to lease the wetlands to wealthy individuals, peasants and entrepreneurs for the purposes of reclamation so as to make the wetlands suitable for agriculture and revenue generation.

I then turn to the second moment in the post-reclamation history of the region when the modalities of colonial gazetteer writing and the promulgation of colonial Forest Act began to produce a different portrayal of the region, a portrayal in which lay the seeds of the delta emerging as a heritage site in the future. The Forest Act did not instantly put an end to reclamation and settlement, but it was a significant historical moment that considerably influenced the status of the Sundarbans after independence. The chapter then turns to the third significant moment in post-independent India when a wide variety of literature on the Sundarbans were produced. I contend that these literature produced at the official and individual levels further consolidated the image of the delta as primarily a wildlife reserve not congenial to human living. Our historical journey in this chapter ends with an account of the Marichjhanpi event in the late 1970s when thousands of refugees settled on a forested island of the Indian Sundarbans were evicted and killed on the ground that they occupied the land meant for tigers.

Colonial rule and the reclamation of the Sundarbans

The Sundarbans is a network of tidal rivers, creeks and islands. The whole region is a delta and exhibits the process of land-making in an unfinished state. According to Richards and Flint, Bengal offers a nearly endless frontier where land continues to be made afresh in the delta (Richards and Flint 1987, 7). The region has been focused on as being largely a forest, but human settlements are as much a part of the landscape of the Sundarbans as the forests. These settlements have been grafted on to forests, which were systematically cleared to make way for human habitation. Eaton sums up the biography of the Sundarbans under colonial rule in the following words:

> At the advent of the British rule in 1765, the Sundarbans forests were double their present size ... During early British rule, zamindars, or landholders, were allowed to continue reclaiming as much of the jungle bordering their plots as they had been doing under the Mughals. In 1828, however, the British assumed the proprietary right to the Sundarbans and, in 1830, began leasing out tracts of the forests... for undertaking the clearing operations preparatory to planting paddy. Then followed forty-five years of rapacious reclamation until 1875–76 when the government declared unleased forest reserved and placed it under the jurisdiction of the Forest Department (Eaton 1987, 1).

Expansion of human settlement into the Sundarbans and reclamation of wetlands accelerated under colonial rule. Reclamation during this period started with a conflict between the colonial state desirous of making inroads into these wetlands for the purposes of revenue earning and the landlords or cultivators exerting their proprietary right on the ground that the wetlands belonged to them.

In the absence of a clear-cut boundary of the Sundarbans wetlands, landlords, in an attempt to extend their cultivated land, continued to 'encroach upon the forest'. The promulgation of the Permanent Settlement Act elsewhere in Bengal in 1793 did not change the Sundarbans scenario for the better. Under this Act, the East India Company vested land ownership in a class of landlords who were required to pay to the colonial government a fixed tax proportionate to the size of their holding. In the event of their inability to pay the required revenue, their land would be taken away and sold to another bidder. The Settlement did not resolve the status of those extensive 'waste' land or jungles, which were not as yet cleared or cultivated in the eighteenth-century Bengal.

A series of regulation acts were passed by the Company to tighten control over reclamation and revenue generation. With this aim in view, Regulation IX of 1816 appointed a special Commissioner for the Sundarbans and vested him with all duties, powers and authority of a Collector of land revenue (Pargiter 1934, 10). Prior to the appointment of the special Commissioner, there were attempts to assess the reclaimed Sundarbans lands, but those attempts remained largely unsuccessful for want of requisite data on the status and progress of reclamation of the Sundarbans wetlands. It was the duty of the Commissioner to define the southern boundary of the districts of 24 Parganas, Nadiya, Jessore, Dacca, Jellalpur and Bakarganj so as to enable the government to fix the limits of the country to be placed under his authority (Ibid). This would require a full and minute investigation. The task involved considerable uncertainty because it was not always easy to know how far the cultivation stretched to the south. The land assessment also proved difficult because in the absence of Permanent Settlement, land not included in the known boundaries of the zamindars would not be liable for assessment. A year later, Regulation XXIII was passed which aimed to define the right of the government to the revenue of the lands not included in the boundaries of estates for which Permanent Settlement was made in the districts of 24 Parganas, Nadiya, Jessore, Dacca, Jellalpur and Bakarganj, which vindicated the government's claim to the 'extensive tracts of land' in the Sundarbans on the grounds that these tracts had been 'waste' at the time of Permanent Settlement in 1793 and, therefore, not included in the Settlement (Ibid, 12). To counter the process of illegal land control, the Commissioner had to embark upon a periodic survey and assessment of the lands mainly in the district of 24 Parganas where such encroachments on the forests were frequent.

Finally, to settle the matter once and for all, the government passed Regulation III of 1828 declaring that, 'The uninhabited tract known by the name of the Sundarbans has ever been, and is hereby declared still to be, the property of the State' (Ibid, 22). The remaining forested part of the Sundarbans would not be assigned to any zamindars or would not be included in any arrangements of perpetual settlement. The Governor-General in Council was empowered to make grants and leases of any part of that Sundarbans and undertake such measures for the clearance and cultivation of the tract as would be deemed proper and necessary. There had been attempts on the part of the government to survey and to define the Sundarbans boundary. This was largely to ensure that revenues could flow from those uninhabited and newly reclaimed parts of the wetlands which were not included in the arrangements of Permanent Settlement. These initiatives finally resulted in the establishment of the demarcating line of the Sundarbans region in 1832–1833 by William Dampier, the Commissioner and Lieutenant Hodges, the Surveyor. As has been stated earlier in the introductory chapter, it is this Dampier-Hodges line that even today remains the authoritative demarcating line of the Sundarbans region.

Driven by the urge to earn more revenue from reclaimed wetlands, the government came up with a new scheme in 1829, whereby lands free of any taxes to the government for twenty years were made available to wealthy individuals who had the means to carry out reclamation. Thus, the government's objective was twofold: to attract new grantees and reduce the burden on the colonial exchequer of reclaiming the remote parts of the Sundarbans wetlands. From 1830 to 1843, 138 such grants measuring an area of 2,022,732 bighas had been given to ninety-five persons interested in reclaiming and turning reclaimed lands into rice fields (Ibid, 57). Reclamation had been carried on by means of woodcutters and coolies procured from Hazaribagh[1] in eastern India (Ibid). In 96 grants inspected by the surveyor, out of an area of 1,422,792 bighas, 201,910 bighas had been brought under cultivation and 28,450 bighas were low jungle, while other grants were all forest (Ibid). The lands granted in the 24 Parganas were surrounded by small creeks supplying freshwater during half the year and obtained enough raiyats, but the lands near the sea did not have those advantages and were liable to inundation (Ibid). On the whole, the reclamation advanced, stood still or receded much according to the individual nature of the grant and the circumstances under which the grantee carried out reclamation and cultivation (Ibid). The promulgation of the Forest Act in the 1870s brought about a change in the colonial approach to reclamation, as the act prevented wholesale leasing of the remaining wetland forests of the 24 Parganas and Khulna.

The reclamation history after the 1870s was replete with frequent changes in the rules of land grant to attract entrepreneurs or industrious peasants to undertake reclamation on behalf of the government. By the turn of the century, it became apparent that reclamation in Bakarganj (now in Bangladesh) was more successful than that in Khulna (now in Bangladesh) and the 24 Parganas (now in India). In 1904, the Sundarbans Commissioner's report on the nature of lands leased in the 24 Parganas showed that only forty per cent of wetlands had been reclaimed in the district (Ascoli 1921, 122). The factor that inhibited extensive reclamation was the 24 Parganas' proximity to the sea and exposure to abnormally high tides. Unlike Bakarganj and Khulna, the area of the delta that forms the 24 Parganas Sundarbans had depended for its development on tidal action alone. Therefore, the delta developed backwards from south to north and the seaward face was more elevated than the interior (Ibid, 158). As a result, big embankments needed to be built to protect the reclaimed parts (the interior), preventing the elevation of ground level by a natural process of silt deposits (Ibid, 121). Thus, landscapes had been engineered and inscribed by technical vision and political exigency (Mosse 2003, 1). The Sundarbans' landscape and hydrology were transformed and shaped by colonial regime of science and commerce. This had brought about significant ecological changes in the region. Thus, from the early stages of reclamation embankments became an issue of considerable significance. In my subsequent discussions, I will show how the question of premature land formation, settlement and embankment building in the 24 Parganas is still a much discussed issue in the post-independent development narrative of the Indian Sundarbans and used by ministers, administrators and developers as a pretext to justify present day human suffering. The reclamation history of the Sundarbans thus moved through various phases, demonstrating endless negotiations between colonial officials on the one hand and landlords, investors and speculators on the other.

Technologies of colonial rule: Gazetteers and statistical accounts

The primary and abiding interest of the colonial government in the agriculture of Bengal, or for that matter anywhere else in India, was the extraction of a part of the surplus in the form of land revenue (Chatterjee 1982, 114). It is equally true that the Bengal forest rules came into force as late as the 1870s, and in the period prior to that the East India Company in general viewed forests chiefly as limiting agriculture (Sivaramakrishnan 1997, 75).[2] However, exploitation and extortion were not the only ways in which the colonial authorities could engage with the landscape and resources of the colonized terrain. Even for the material needs of colonial rule, complete and exhaustive knowledge about the land and its people was necessary. The colonial 'statemaking' could successfully proceed only when the country was made available for knowledge through representations. To this end, there emerged a new genre of writing in the form of gazetteers, statistical surveys and censuses. The second significant historical moment in the biography of the Sundarbans arrived when the delta like many other regions in India became amenable to gazetteering and enumeration. It was in this representation as found in the gazetteers of Hunter

(1875) or O'Malley (1914) that lay the prospect of the Sundarbans emerging as a World Heritage Site, a site to whose conservation inhabitants became the obvious obstacles.

Here my focus on gazetteering or gazetteer writing can be situated in a broader context of a trend towards conceptualizing the epistemological dimensions of colonial rule in India (Chatterjee 1995; Bayly 1996; Cohn 1997; Dirks 2001). Colonial rule, as is more or less widely acknowledged, was sustained not merely by superior arms and brute force, but by constant attempts on the part of the colonizer to define and redefine an epistemological space for the colonized. Colonial rule, to be effective, relied on a complete and exhaustive knowledge about the society colonized. Not only did it entail a construction of India's pre-colonial past, but also an exhaustive survey of its colonial present. Cohn (1997) highlights the significance of surveys, gazetteers and censuses in connection with his discussion of the different 'modalities' of knowledge building in colonial India. Surveys and gazetteers as 'enumerative modalities' aimed at classifications of the land and its people and pigeonholing them on the basis of what the 'technologies of colonial rule' (Dirks 1997, 2001) prescribed as 'scientific' criteria. 'India was an ideal *locus* for science' (Prakash 1992, 155; italics as in original); the rich diversity, which India offered was to be mined for knowledge and the convenient tool that aided this enterprise of science was one of classification. Thus, the objective was not simply to codify law and bureaucratize governance, but to survey, classify and enumerate scientifically the geographical, geological, botanical and zoological properties of the natural environment and the archaeological, demographic, anthropological, linguistic and economic traits of the people (Chatterjee 1995; Grout 1996).

On the question of nature of colonial knowledge production, Bayly differs from others and registers his sense of discomfort with the term knowledge, for assimilation of power into knowledge, according to him, makes it difficult to analyse the systems of communication and surveillance, or the gaps and contradictions within them (Bayly 1996, 366). Instead, he uses the term 'information' to show how the information order of the British, far from constituting a complete epistemological break with the pre-colonial order, actually relied heavily on it. However, what all the above authors agree, despite differences in their orientations, is the fact that after the 1857 rebellion, there occurred a shift in the nature of colonial knowledge from textual to empirical and functional. The mutiny made it clear to the British that their knowledge about the colonized population was not only empirically inadequate, but far too unsystematic (Bayly 1996; Cohn 1997; Dirks 2001). In the wake of gazetteers and censuses as authentic institutional procedures, information collected about the land and its people developed into an ever-increasing pool of knowledge, disembodied, objective and carefully documented for future reference, transmission and reproduction. Our discussion of gazetteers and surveys as modalities of knowledge production forms a backdrop against which I intend to trace briefly the history of how the Sundarbans was taxonomized into a world resource site. Here we are concerned with gazetteers not simply as instruments of colonial rule, but also as important vehicles for transmitting and reproducing the dominant image of the region in postcolonial India.

Colonial knowledge building and the Sundarbans wasteland

The decade of the 1870s saw the publication of a series of gazetteers and statistical accounts of the areas, which came to be organized into districts. Such a task involved surveying topography, natural history, antiquities, taxation, local customs, diet and general living conditions. W. W. Hunter, a Bengal cadre of the I.C.S., responsible for compiling and surveying these district-level operations, in his preface to Volume I of the *Statistical Account of Bengal* quotes the Court of Directors as saying:

> "We are of the opinion," wrote the Court of Directors in 1807 to their servants in Bengal, "that a Statistical Survey of the country would be attended with much utility: we therefore recommend proper steps to be taken for the execution of the same" (Hunter 1998 [1875], xxiii).

Several years passed before the task of compiling the statistical survey of the districts was finally entrusted in 1869 to Hunter. Two years later, in 1871 Hunter became Director-General of Statistics and by 1875 his *Statistical Account of Bengal Volume I* (District of 24 Parganas and the Sundarbans) appeared along with several statistical accounts of different provinces in India. By 1881, the compilation of all the provinces was completed, and this was followed by the publication of *The Imperial Gazetteer* in 1881 and *The Indian Empire: Its History, People and Products* in 1882.

The reason why I have touched on the colonial project of gazetteer writing is to show that the Sundarbans, just like other parts of the colonized land, was made available for statistical account and enumeration. As already mentioned, the objective behind such statistical accounts was to gain exhaustive knowledge about the land and deploy it for the purposes of colonial rule. Hunter had scrupulous regard for facts, and the purpose of his statistical account was to produce exact local knowledge for administration of Bengal (Greenough 1998, 238). In realizing this objective, Hunter produced his first volume on the 24 Parganas and the Sundarbans. In attempting a statistical account of the Sundarbans, Hunter tried to resolve the major dilemma that confronted the colonial administrators at various stages of the Sundarbans reclamation, namely, whether to convert the wasteland to a cropland through reclamation and settlement or deny people's claims to this land by declaring it a forested wasteland and, therefore, the state property.

For Hunter, the Sundarbans was a drowned land, full of jungles and an abode of wild beasts. In short, he portrayed the delta as a sodden wasteland (Ibid, 240). We already noted that from the beginning the Sundarbans emerged as a wasteland over which conflict arose between the colonial state and the zamindars. By declaring the Sundarbans a wasteland, Hunter placed renewed emphasis on the concept of 'waste'. This emphasis was placed in the mid-1870s when the terminology of 'waste' was being readied by colonial policymakers to deflect native claims to the vast forests of India in order to move towards their exploitation (Guha 1990). In 1878, the Forest Act was promulgated and the heavily wooded Sundarbans was also brought under the purview of the Forest Act (Greenough 1998, 240). The Forest Act declared the unleased forest of the Sundarbans reserved and slowed down considerably the process of leasing out of forested lands for reclamation and agriculture. Thus, with the promulgation of the Forest Act, the twin principles of exploitation and protection started, exploitation for the purposes of colonial rule and protection of forests from the natives on the grounds that they did not know what scientific forestry was all about. Therefore, it is not surprising that the Sundarbans was depicted as a sodden wasteland and such depiction fitted well into the colonial project of gazetteer making and was geared towards the utilitarian purposes of colonial rule (Ibid).

However, the question that remains is why Hunter chose to devote such attention to a place he characterized as sodden wasteland. The statistical account adopted the district as the unit of description. This was more or less followed everywhere in Bengal. But when Hunter turned to the Sundarbans, he abandoned the district of 24 Parganas in favour of the geographic region (Ibid, 247). The low, tide-washed archipelago crossing the sea face of the 24 Parganas, Jessore and Bakarganj refused to be shoe-horned into a schema designed for the 225 districts of British India (Ibid). Following Greenough, one could ask why Hunter chose to treat the Sundarbans as a single unit instead of incorporating it piecemeal into essays on the adjacent districts of 24 Parganas, Jessore and

Bakarganj (Ibid). Hunter's intention was to write a readable account and an administrative manual. Beneath the overt statistical account with its fixed topical format of district account lurks another intuitively assembled version of the Sundarbans (Ibid, 248). There was more than the utilitarian need of gazetteer writing which was driving Hunter's account. According to Greenough, it was Hunter's Victorian sensibilities that led him to engage with the landscape and geography of the Sundarbans (Ibid, 240). To quote Hunter, 'The southern portion of the Sundarbans, which comprises the jungle tract along the seashore, is entirely uninhabited, with the exception of few wandering gangs of woodcutters and fishermen. The whole population is insignificant' (Hunter 1998 [1875], 35). Thus, for Hunter, the humans were all 'immigrants' and the tigers and crocodiles were the only 'aboriginals' (Greenough, 247). Even when Hunter proceeded to catalogue human groups resident in the reclaimed areas, they appeared only after a longer list of snakes, birds and fish had been presented (Hunter 1998 [1875], 15). On the arduous task of reclaiming the land, Hunter commented:

> So great is the evil fertility of the soil, that reclaimed land neglected for a single year will present to the next year's cultivator a forest of reeds (*nal*). He may cut it and burn it down, but it will spring up again almost as thick as ever (Ibid, 52; italics as in original).

For Hunter, the soil was fertile, but so awfully fertile that it was suitable for anything but cultivation and therefore, human habitation. Hunter's portrayal was equally powerful when it came to describing nature's fury:

> … the inundation works cruel havoc among [the] low-lying isolated villages. The grain in their fields is spoiled; their houses are torn away … Liability to cyclones must put a practical limit to the extension of cultivation … the more the forest is cleared away, the smaller the barrier placed between the cultivator and the devouring wave (Ibid, 55–56).

In other words, Hunter seemed to be arguing that in a shifting terrain like this, it was people who remained so to speak 'out of place', thus legitimizing the attention he gave to wildlife. Where Hunter talked about nature the first thing that caught his attention was the tiger.

> Tigers are very numerous, and their ravages form one of the obstacles to the extension of cultivation… The depredations of a single fierce tiger have frequently forced an advanced colony of clearers to abandon their land, and allow it to relapse into jungle (Ibid, 33).

Hunter did not mention in so many words that the Sundarbans was totally uninhabitable for humans, but his portrayal suggests that it was not very suitable. A few years later O'Malley described the Sundarbans as a region of total desolation where there was nothing to induce an influx of immigrants, and where even the fecundity of the inhabitants seemed to be sapped by endemic fevers and epidemic diseases (O'Malley 1913, 26). A year later when O'Malley wrote his 24 Parganas District Gazetteer he opened his account with an exhaustive classification of the botanical and wildlife resources, assuming that a gazetteer of the 24 Parganas must start with a description of the flora and fauna of the wondrous region called the Sundarbans (O'Malley 1998 [1914]).

The Sundarbans in post-independent India

To present Hunter's account or O'Malley's views is not to suggest that they decided the status of the Sundarbans in definitive terms. However, in these portrayals lay the prospect of the emergence of the Sundarbans as a distinctive place in post-colonial India. In 1973, the Sundarbans forest was declared a tiger reserve because it is the only mangrove tiger land in the world. In 1984, the Sundarbans became a National Park. Soon after that in 1989, the Sundarbans was declared a Biosphere Reserve in which large stretches of mangrove forest, containing sixty-four mangrove species, the highest in a single area, had been conserved and wilderness maintained with its original ecosystem intact under the protective shelter of Project Tiger (Directorate of Forests n.d., 3). In the same year, the Sundarbans was declared a World Heritage Site for its unique ecological endowments. Thus, the delta, which started its career as a wasteland under the colonial regime of reclamation, found its way onto the widely acclaimed list of World Heritage Sites.

If the decade of the eighties saw the Sundarbans attaining worldwide fame, the decade of the nineties witnessed the reprinting of colonial documents such as Hunter's statistical account and O'Malley's District Gazetteer by the Government of West Bengal. When the first volume of Hunter's statistical account of Bengal was reprinted in 1998, the section on the Sundarbans was reprinted separately. O'Malley's gazetteer was reprinted in the same year with the picture of tigers on the cover page indicating the importance of the tiger when talking about the district of 24 Parganas. When the West Bengal Government's Information and Cultural Department published its volume on the district of South 24 Parganas a tiger not only appeared on the cover page, but on virtually the first page of each article. The volume contains articles that discuss various aspects of the district, but the common theme that runs through them is the Sundarbans, its wild plant and animal life and the prospect of its conservation (West Bengal 2000).

Literature on the Sundarbans' ecology and tigers has been produced both at official and individual levels. Rathindranath De, the Forest and Tourism Secretary of the Government of West Bengal, promoting the Sundarbans to tourists, writes that the shimmering tidal waters bordered by mangrove trees are like a world of fantasy. The visitor suspends normal time and embarks on a slow and lazy cruise against the tide along estuaries (De 1990, 1). In keeping with the needs of ecotourism, a powerful concept in recent times, the Tourism Department has drawn up programmes involving the setting up of an interpretation centre at Sajnekhali Reserve Forest, a turtle breeding centre at Bhagabatpur and Bakkhali and watchtowers to view tigers at various places in the region (Directorate of Forests n.d., 18). Hunter's drowned land has thus been converted to suit the needs of tourism, an important source of revenue for the government. Others writing on the Sundarbans have also come up with equally powerful descriptions. Haraprasad Chattopadhyay in his *Mystery of the Sundarbans* describes the place as a peculiar terrain where the man-eating Royal Bengal Tigers live as the jealous neighbours of the man-eating ferocious crocodiles, poisonous snakes, lizards, birds, sharks, honeybees etc. Both the fury and beauty of the place have attracted the attention of tourists, ecologists and zoologists (Chattopadhyay 1999, 6).

Similar portrayals can also be found in the writings of historians (Das et al. 1981) and geographers (Guha and Biswas 1991; Banerjee 1998). Vernacular literature available on the Sundarbans has described the place in a more or less similar vein. However, this is not to suggest that writers have concentrated only on nature and natural beauty of the delta. Issues discussed include patterns of settlement, nature of population (Das et al. 1981; Guha and Biswas 1991; Banerjee 1998), history of reclamation (Das et al. 1981; Mondal 1995; Mondal 1997) and folklore and archaeological remains of the delta (Mondal 1999; Jalil 2000). Writers have also looked at the economic life and poverty of the people (Das et al. 1981; Mondal 1995). The problem with the approaches adopted to the study of the Sundarbans is that writers, in reflecting on the Sundarbans, have tended to treat the image of the region as an abode of wildlife where human beings are secondary as the natural fact and have not carefully analysed how this natural fact has been socially constructed.

In her book *Spell of the Tiger*, Montgomery describes her visit to the Sundarbans in the following manner:

> On the wide rivers you may see Gangetic dolphins rise, smooth as silk, their pink-gray dorsal fins rolling like soft waves along the water's surface. Dreamlike wonders: once, out near the Bay of Bengal, I glimpsed an olive ridley sea turtle as it surfaced for a breath of air ... "But once you put your foot on the mud bank, you know: this is a strange place ... a dangerous place ..." ... where you know, for the first time, that your body is made of meat (Montgomery 1995, 2–3).

The above paragraph conveys a powerful depiction of the region. Where else on earth would one be made to feel as though one's body was made of meat except in a place so strange that it was not habitable for humans? Montgomery provides a graphic description of a place where the tiger reigns supreme. But, what prompts her to write the book is the apprehension that the tiger is endangered. Her book, she believes, is an invitation to visit the land, a journey, she fears, people may soon be unable to make, for there may be no more tigers to attract people there. Montgomery refers to the poaching of tigers and overcrowding of the forest. It is the same apprehension that has led the state government to launch a programme for the protection of the tiger with funding from international conservation groups such as the World Wildlife Fund [WWF] (Mallick 1999). However, all these perpetuate the image that the tiger is of primary importance and that people are secondary. Although the Sundarbans Biosphere Reserve was launched to strike a proper balance between the human and non-human inhabitants of what is considered a unique ecosystem, this balance is tilted in favour of the tiger.

When people entering the forest or creeks get killed by a tiger, it is justified by officials through recourse to the argument that the people have been 'intruders'. But when a tiger strays into inhabited islands, killing humans and livestock, it is believed the animal is hungry. Time and again concerns are raised over how starvation is causing the death of these supreme predators (Chaudhuri 2002). Thus, people are portrayed as unauthorized occupants of a land whose exploitation of the forest resources endangers tigers. Catching tiger prawn seeds in the river causes depletion of marine resources. So the very presence of humans on the islands is a menace to the future conservation of the forested delta. A volume entitled *Wilderness: Earth's Last Wild Places*, published by a team of more than 200 international scientists, has identified the Sundarbans as one among thirty-seven of the earth's most pristine areas critical to the earth's survival and where over-exploitation of resources and human settlement are seen to be posing a threat to the place (*Hindustan Times* 2002).

The above discussion leads one to think of 'displacement as a conservation tool' (Sharma and Kabra 2007, 21). To ensure the conservation of a nationalist project and a site for tourism, people settled on the site need to be kept at bay. In their research on the Kuno Wildlife Sanctuary in Madhya Pradesh, the authors draw attention to the Wildlife Institute of India's recommendation in favour of a relocation of the villages to reduce the probability of human-lion conflict (Ibid, 28). Accordingly, a rehabilitation package was offered to the villagers whereby the villages were relocated to a place away from the sanctuary. Sharma and Kabra note that this relocation had proved costly for the people because at the new site they were forced to adopt agriculture-based livelihood with which they were not familiar. This was a community of hunters and gatherers who suddenly found themselves reduced to the status of wage labourers, who increasingly found themselves engaged in various construction activities commissioned at the relocation site. Like other rehabilitations, the conservation-induced relocation is fraught with its own set of problems. Johari voices similar anxiety in her research on the tiger conservation project at Sariska (Johari 2007). A national park based on the idea of excluding people can prove to be disastrous as has been the case with Sariska where poaching of tigers, inflated tiger census, and complicity of forest staff in poaching activities have involved forest authorities and local people in a mutual blame game. The Sundarbans wildlife sanctuary presents us with a vision of conservation where displacement works through the principle of denial, that is, denying people an access to resources of the region.

The Marichjhanpi event: Wilderness consolidated

Nowhere was this conservation cum development imperative more evident than in the Left-front government's policy towards the East Bengali refugees who settled on Marichjhanpi island of the Sundarbans. Ever since the partition of India in 1947, East Bengali refugee rehabilitation had been an issue that confronted the Government of India. The refugees from the middle and upper classes had managed resources and built connections to start a new life in West Bengal without being primarily dependent on government assistance (Mallick 1993, 97). However, poor East Bengali refugees belonging largely to untouchable castes who were dependent on government assistance were settled by the central government in Dandakaranya, a place that is part of

Orissa and Madhya Pradesh (Ibid, 99). Before the Communist Party (CPI-M) came to power in West Bengal, they demanded that refugees in Dandakaranya be settled in the uninhabited Sundarbans delta in West Bengal (Ibid, 99). In fact, the refugees were given to understand that once the party came to power, they would be settled in Bengal, a place where they would feel at home. This enabled the communists to obtain a political following among these refugees settled outside West Bengal. During the B. C. Roy government of the 1950s and early 1960s, Jyoti Basu, the then opposition leader, presented their case in the Legislative Assembly and demanded later in a public meeting in 1974 that the Dandakaranya refugees be settled in the Sundarbans (Ibid, 99). In a letter dated 13 July 1961 to the Minister of Refugee, Relief and Rehabilitation, Jyoti Basu stated,

> Prolonged hunger-strike by the refugees lasting for more than a month in almost all camps in West Bengal has proved beyond doubt strong reluctance on the part of the refugees to accept proposal of the Government regarding their rehabilitation in Dandakaranya ... We do not think that the rehabilitation of the camp refugees in manner acceptable to them is so very difficult as is often being suggested by the Government ... many have found rehabilitation in West Bengal although it was stated by the Government that West Bengal has reached a saturation point. I feel, therefore, that the rest may be found rehabilitation here provided there is willingness on the part of the Govt. The enthusiasm that will be generated among the refugees if such a policy is accepted will be no mean an asset for their proper rehabilitation (Mondal 2002, 37–38).

At a meeting of the CPI-M and its allies in 1975, it was resolved that the refugees be settled in the Sundarbans and a memorandum to that effect was proposed to be submitted to the Governor (Mallick 1993, 99–100).

In 1977, when the Left-front came to power, they found that the refugees had taken them at their word and in 1978, some 150,000 refugees arrived from Dandakaranya (Ibid, 100). The Left-front government saw these refugees as obstacles to the economic recovery of the state. The last thing that the Left-front government wanted was a refugee influx which might damage the prospects of an economic recovery in the state and divert scarce resources from other development projects (Ibid). The communists, therefore, went back on their earlier policy and most of these refugees were detained in the camps set up by the communist government and forcibly sent back to Dandakaranya. However, in May 1978, about 30,000 refugees managed to cross the riverine delta area and settle in Marichjhanpi, an island lying to the northernmost forested part of the West Bengal Sundarbans. Within a short span of time, the settlers set up a viable fishing industry, health centre and schools (Ibid). However, the state government was not disposed to tolerate this. The Chief Minister, Jyoti Basu, who had once as opposition leader defended the refugees' case, now declared that the occupation of Marichjhanpi was an illegal encroachment on Reserve Forest land, on the state and on the tiger protection project sponsored by the World Wildlife Fund (Mallick 1999, 115). The Chief Minister further stated that if the refugees did not stop cutting trees the government would take 'strong' action. He had said, 'Enough is enough, they have gone too far' (Chatterjee 1992, 298–299). When persuasion failed to make the refugees abandon their settlement, the government started an economic blockade of the settlement (Mallick 1999, 107–108). Thirty police launches were deployed to cut off their supplies. Their huts were razed, their fisheries and tube-wells destroyed. When the settlers tried crossing the river for food and water their boats were sunk. To clear the island, the police opened fire killing thirty-six people. Forty-three more died of starvation, twenty-nine from disease and 128 from drowning when their boats were sunk by the police (Mallick 1993, 101).[3] Many families who survived and were forced to leave Marichjhanpi for Dandakaranya perished in transit due to starvation and exhaustion.

The refugees were denied settlement and evicted because they occupied land meant for tigers. Mallick notes that there appears to be nothing on record indicating any pressure on the government for eviction from any environmental non-governmental organizations or non-state groups (Mallick 1999, 115). Marichjhanpi proves that people's needs are viewed as obstacles to Sundarbans development. During my fieldwork, I found that Marichjhanpi still survived in the public memory and the episode often came up in the course of my conversation with the villagers. According to them, the Marichjhanpi incident was a reminder that in the Sundarbans, the tiger comes before humans. Reflecting on the killing of the refugees a villager once sarcastically asked me, 'So many human bodies in the rivers and creeks, it must have been a good treat for the tigers, what do you think?' His sarcasm was shared by many others who felt that living in the Sundarbans put them at the mercy of the tigers.

Both Mallick and Chatterjee emphasized upon the caste dimension of the Marichjhanpi incident. According to them, what was significant to note here was the identity of the East Bengali refugees as untouchables. The communist government, which orchestrated the Marichjhanpi episode, had many East Bengali refugees constituting the upper layer of the party leadership. Some of the top ranking cabinet members including Chief Minister Jyoti Basu in the Left-front government were from East Bengal. However, these leaders were settled educated upper class/caste East Bengalis. In their eyes, the untouchable refugees were dispensable and their interests were non-negotiable. Therefore, the party, which once championed the refugees' cause, had no qualms in betraying them when it came to power in West Bengal. Jalais' article (2005, 1757) suggests how the government's primacy on ecology and its use of force in Marichjhanpi was seen by the Sundarbans islanders as a betrayal not only of the refugees and of the poor and marginalized in general, but also of the Bengali backward caste identity. For the Sundarbans islanders, Marichjhanpi was a betrayal by the government and the Kolkata urbanites (Ibid, 1761). Mondal's (2002) *Marichjhanpi: Noishabder Antaraley* (Marichjhanpi: Hidden Behind the Veiled Silence) presented the detailed lists of people killed in the violence perpetrated by the state.[4] Subsequently, Marichjhanpi became a subject matter of vernacular literature in West Bengal. Pal's book (2009) *Marichjhanpi: Chhinna Desh, Chhinna Itihas* (Marichjhanpi: Fragmented Homeland, Fragmented History) presented a well-documented narrative of the incident. Pal's book contained valuable newspaper reports highlighting the massacre and also West Bengal Legislative Assembly debates between the left Chief Minister Jyoti Basu and the SUCI (Socialist Unity Centre of India) opposition leader Debaprasad Sarkar. Ghosh (2013) argues that for long Marichjhanpi lay hidden[5] in the archive as a forgotten chapter in the history of left rule in West Bengal. It resurfaced only in the context of protest movements in Singur and Nandigram[6] serving to remind us of the irresistible fascist tendencies in the communist rule in Bengal. Pal's book[7] revisited that long forgotten moment when the state executed violence in curbing the aspirations of the dalit refugees in search of their homeland.

However, Marichjhanpi in this chapter appears as a significant moment in the consolidation of the Sundarbans' image as a natural wilderness. Guha and Martinez-Alier (2000, 23)[8] argue that Marxism's relation with environmentalism is somewhat uncomfortable, as most Marxists chose to interpret environmentalism as a frivolous upper class fad; as dangerous, romantic and anti-industrial, ignoring the interests of the poor. Marxists have resisted the introduction of ecology into historical explanation perhaps because of the fear that this could naturalise human history (Ibid, 26). Historically, the communists were engaged in various land-based struggles and agrarian movements in colonial and post-independent Bengal. The left parties such as the CPI-M or RSP were also involved in land movements in different parts of the Sundarbans. The same left parties, who championed the rights of the landless and poor, placed primacy on ecology when they turned to the refugee settlement in the Sundarbans. Interestingly, the communists became environmentalists, upholding the so-called upper class fad much to the neglect of the refugees' cause. The Marichjhanpi incident did not evoke any negative response or protest in the larger Bengali society because a much stronger portrayal of the Sundarbans as primarily a wildlife reserve had already gained considerable ground. This image of the Sundarbans came in handy in the Left-front's denial to the refugees their right to settlement in the Sundarbans. Human settlements continue in the Sundarbans even after the Marichjhanpi massacre. But Hunter's sodden wasteland was destined to become a World Heritage Site in the 1980s. Marichjhanpi remains an important moment, a significant link in the chain of events leading the Sundarbans to attain worldwide fame.

Powerful portrayal and missing links

It was to reflect on the biography of the Sundarbans that a conference entitled 'The Commons in South Asia: Societal Pressure and Environmental Integrity in the Sundarbans' was held in 1987 at the Smithsonian Institution in Washington, DC. Although the focus was mainly on the Bangladesh Sundarbans, a number of papers presented dealt with the pre-partitioned Sundarbans of Bengal and its history of land reclamation. Some of the papers presented have been cited earlier in this chapter when the history of land reclamation was narrated. According to Herring, one of the contributors to the conference, the central dilemma in the Sundarbans development is that, unlike the tribal forests elsewhere in South Asia, where the conflict is between the utilization of an existing habitat cum common-property resource and a historically novel statist claims to management, the remaining and shrinking mangrove forests have become an object of conflict between social forces seeking a livelihood and a state that seeks to limit that process (Herring 1987, 10). Herring instructs us to revisit the biography of the Sundarbans during colonial times when the practicalities of colonial statemaking went hand in hand with the representations made of the region. It is this representation of the place as a natural wilderness that has implications for the way development of the region has been perceived in post-independent India and explains why the view that the Sundarbans can be best developed if left to grow without hindrance has gained precedence. It is this development imperative that conceives of the place as suitable for tourists, visitors, botanists and zoologists, but not for its inhabitants, who are considered to be intruders.

Herring reminds us of the broad imperative underlying the Sundarbans development. For Hunter and O'Malley, the Sundarbans was a wasteland and a place full of diseases. The passages from Hunter's Statistical Account upheld the region's beautiful natural landscape, but only with a caveat that the place was so dangerously beautiful that it was inhospitable. In this portrayal lay the prospect of the emergence of the Sundarbans as a distinctive place in post-independent India. Marichjhanpi and the regime of gazetteer reprinting were the moments when powerful portrayal became portrayal of power. Such a portrayal deployed considerable power in controlling our imagination about the region and making us think of its development in a particular way. A wide variety of literature produced on the Sundarbans (which I discussed earlier in this chapter) in post-independent India produced and reproduced the dominant portrayal as the 'natural fact' of the region. The result has been a conservation drive based upon the recognition of the tiger as the legitimate claimant to the land and people as only intruders or mere 'tiger food'[9] (Jalais 2010, 11).

However, to say this is not to argue against conservation. My intention here is to situate the imperative of conservation in a historical perspective analysing the missing links between wilderness and history and conservation and politics. Conservation projects have always been expressions of nationalist aspirations. A national park whether in the Sundarbans or elsewhere in India is an instance of 'monumental nationalism' (Rangarajan and Shahabuddin 2007, 6). Statist conservation in India is marked by a lack of sensitivity to social complexities and cultural histories. Management plans for protected areas or tiger reserve never make any passing references to the traditional knowledge, culture or resource uses of local populations, which are treated merely as unwanted elements to be ignored or relocated (Ibid, 10). The discussion above focused upon a historical understanding of the Sundarbans, but history in this chapter did not emerge merely as background information, but as constituter of ideas enabling us to track the missing links, links that help explain why it is important to turn to people's lives in the delta. With this aim in mind, I now move to the third chapter which turns to embankments and governance in the Sundarbans.

Notes

[1] Hazaribagh is situated in what today is the newly formed state of Jharkhand in India. Tribals in the form of coolies were also hired from the Santhal Parganas. Both Hazaribagh and Santhal Parganas are parts of the Chotanagpur Plateau, a part of the erstwhile state of Bihar, which has one of the largest tribal populations in the country.

[2] 'It is the conquest of the forest by arable, of nature by culture'. Culture here implies settled agriculture, a civilizational process, which takes place through reclaiming land from the forests. For details, see Rangarajan and Sivaramakrishnan (2012, 1).

[3] For further discussion of the Marichjhanpi incident see Chatterjee (1992) *Midnight's Unwanted Children*, pp. 291–379 and also Mallick (1993) *Development Policy of a Communist Government*, pp. 97–103, and (1999) 'Refugee Resettlement in Forest Reserves: West Bengal Policy Reversal and the Marichjhapi Massacre', pp. 104–125.

[4] Mondal's book provides detailed lists of people killed in police firing, people who died of starvation, people who died after consuming inedibles, people who went missing, people put behind bars, women raped by police forces. For details see Mondal (2002).

[5] Marichjhanpi lay hidden under more prominent success stories of the left governance in Bengal. Left political success largely revolved around two experiments executed in rural Bengal, namely land reform and operation barga and panchayat institutions of self-governance and decentralization. During thirty-four years of left rule literature produced on West Bengal have largely focused on the above political programmes launched by the Left-front government.

[6] There were protest movements in Singur (Hooghly district) and Nandigram (Purba Medinipur district) against forcible land acquisitions for the industrial policy pursued by the left. Singur and Nandigram have been almost epochal and iconic events, which played a very significant role in dislodging the Left-front and facilitating its ouster from power.

[7] Pal's (2009) book also has a rare collection of essays offering diverse perspectives on the event. The book has incorporated an essay by Amiya Kumar Samanta, the Superintendent of Police, who orchestrated the violence deploying police launches to execute the economic blockade. According to Samanta, the Marichjhanpi incident was blown out of proportion by politically motivated people. The refugees were ready to leave the island and they left on their own. They were not forcibly evicted from the island. Marichjhanpi was sensationalized by vested interests.

[8] Guha and Martinez-Alier argue that following their western counterparts, Southern Marxists rejected environmentalist critiques because they thought it was a western ploy to keep the Third World underdeveloped. The authors trace how the Marxists' neglect of ecology goes back to Marx and Engel's own negative reaction to Sergei Podolinsky's attempt in 1880 to introduce human ecological energetics into Marxist economics. Authors view Podolinsky's analysis of energetics of life and its application to the analysis of economy as highly significant. The decades of neglect of the study of energy flow by Marxist historiography and economics have continued to this day. For details see Guha and Martinez-Alier (2000).

[9] Jalais elaborates this point in her article (2005) and also subsequently in her book (2010). She argues that Marichjhanpi has been orchestrated by the Left-front government and in this, the government had been supported by the bhadralok of Kolkata (the educated upper caste/class people which included among others the East Bengali refugees settled in Kolkata as well), who were instrumental in Left-front's rise to power in 1977. For these urbanites, the refugees, belonging to the so-called lower rung of the

society, were 'tiger food' disposable people who could be shot and killed because they wanted a homestead they had been promised (2005, 1760). The villagers felt that the government was happy as long as the tigers thrived, and that in contrast, whether the islanders lived or died, as had been the case for the refugees, made no difference, because they were just 'tiger food' (Ibid, 1761). With the forest increasingly becoming a state property the tiger has become the citizen and the islanders merely a prey for the tiger. For details see Jalais (2005, 1757–62) and also (2010, 146–75).

3

Governing the Sundarbans Embankments Today
Between Policies and Practices

Grand development and islanders' worries

In 1996, the Directorate of Forests of the Government of West Bengal organized in Kolkata a seminar in memory of William Roxburgh who was not only the father of modern botany in India but also the first botanist to have made plant collections from the Sundarbans. In 1999, the proceedings of the seminar were published in the form of a volume called *Sundarbans Mangal*.[1] Kiranmoy Nanda, the Minister-in-charge of Fisheries, Government of West Bengal, spelt out the purpose of this specialists' meeting in the following manner:

> A million dollar question [that] peeps in our minds [is] how Sundarbans is gradually losing her treasures. The Sundarbans, largest delta of the World, is the much talked about natural resource site and it is a privilege for the Indians to have such wonderful place of natural grooming and wildlife habitat. With the increasing population pressure the ecosystem of the Sundarbans is losing its balance slowly. The number of wildlives … have gone down considerably owing to deforestation and the destruction of natural habitats. The occasional visit of "Royal Tigers" to the adjoining villages in search of food is a proof of the above statement (Nanda 1999a, 10).

The seminar provided an opportunity for botanists, zoologists, geologists and ecologists to update their knowledge. The issues debated ranged from ecological balance to sustainable mushroom spawn cultivation. Specialists even dwelt at length on the technicalities of embankment construction and its protection against the tidal waves. This was clearly an attempt by the specialists to make the Sundarbans part of the global discourses of conservation and sustainable development. In 2001, the Sundarbans Biosphere Reserve became part of UNESCO's world network of biosphere reserves and as a result, the state forest department now expects an initial grant of 3 million dollars from the UN Development Programme (UNDP) for the protection of the Sundarbans (*The Statesman* 2002). Today, the World Heritage Site also attracts funding of Rs. 3 crores[2] from the Asian Development Bank (ADB) to conduct research on the development of infrastructure, soil conservation and a livelihood programme for the people in the area. The Sundarban Affairs Department (SAD) will collaborate with the ADB on this. Thus, deals are struck and development agendas are prepared in the government departments. If we recall, the very first chapter of the book opened with the accounts of Bhagirath, Tapan and Chandan, the islanders who expressed their vulnerability to frequent embankment collapse and land loss. They narrated how they were fast losing the ground beneath their feet. Yet, development agendas such as the ones struck in the corridors of power exclude the worries and anxieties of Bhagirath, Tapan and Chandan and several others because they are not deemed to be worth the attention of the community of scientists, developers and policymakers.

Against this background, the chapter attempts to interrogate the rhetoric and practices of the two main government departments that are concerned with the Sundarbans development, namely Sundarban Affairs and Irrigation and Waterways (IWW). I will also endeavour to examine how these two departments have addressed the problems of agriculture and embankments, respectively. To this end, I will analyse documents such as departmental budget speeches, annual reports and utilize interviews with ministers, departmental officials and bureaucrats on the one hand, and focus on the practices of local-level engineers, field officials and functionaries on the other. I will also supplement this with accounts of local people's perceptions about the activities of these two government departments in the Sundarbans. Thus, while this chapter is largely concerned with the state, it does not examine it at the level of policy or programme alone. Apart from looking at various official documents and policy resolutions, I attempt an ethnography of the state's functions at the local level and show how the state is manifested in the activities of its lower or field-level officials.

There has emerged a vast body of literature approaching the state from the point of view of how it functions in the wider society (Handelman 1981; Abrams 1988; Brow 1988; Nugent 1994; Gupta 1995; Fuller and Bénéï 2001). The state is no longer viewed from the point of view of the oppositional model of state-society relations, but from a perspective that encompasses dimensions of cooperation and conflict in state-society relations (Nugent 1994). Breman (2000), in his study of state policies for the welfare of the rural proletariat in western India, has provided a rich and insightful ethnography of the activities of the government labour officer charged with surveying and documenting cases of underpayment among the rural labour force. The officer's dealings with the villagers – employed as agricultural labourers by big landholders – during his visit to the surveyed areas offer an insight into how the state operates through its officials and how its presence is felt in the wider society.

In a more or less similar vein, Ruud (2001) studies the role of politicians of the ruling CPI-M party in a village in West Bengal. These political leaders, according to Ruud, are the representatives of the state and their activities demonstrate the working of the state at the village level. By providing an ethnographic account of a village panchayat meeting convened to elect a village committee, Ruud shows how the local leaders use their clients and cadres to get the consent of the villagers, a consent that is of crucial importance to the legitimacy of the leftist state. Tarlo (2001) reconstructs the Indian state during the period of national emergency between 1975 and 1977 by focusing on more than 3,000 personal files at a municipal office in Delhi and points to the banal bureaucratic workings of the state at the local level. She shows how people who were victims of slum clearance secured their housing rights by undergoing sterilization or motivating someone else to do so. By uncovering this forgotten story of national emergency, Tarlo tells us in minute detail how the clerks of the local municipal office manipulated records and concealed facts about slum clearance. By highlighting the everyday practices of the lower-level officials and functionaries, the writers have drawn attention to the ways in which the state is made visible at the local level. It is this body of literature that will provide the basis for my understanding of the state departments' approach to the Sundarbans development. The chapter also turns to the cyclone Aila and its aftermath in the region. The cyclone struck the Sundarbans at a time when West Bengal was undergoing political transformation. The left rule in West Bengal was coming to an end. With the Trinamul Congress under Mamata Banerjee's leadership turning out to be a major contender of power, the narrative of governmental relief for the cyclone victims was also a story of parties using resources to mobilize their respective support bases. The chapter focuses on Aila relief distribution in so far as it provides insight into how politics was played out at the local level. Finally, I highlight irrigation department's post-Aila policies towards the Sundarbans embankments. And I argue that despite the post-Aila policy articulations concerning embankment reconstruction, land acquisition and compensation, development apathy continues in the Sundarbans.

Sundarban Development Board: The story of an initiative nipped in the bud

The Sundarban Development Board[3] (SDB), which is currently part of the SAD was constituted in 1973 'for integrated and accelerated development' of the Sundarbans region. The functions entrusted to the Board at the time of its constitution were given as follows:

- Formulation of integrated programme for effective utilization of local resources;
- Coordination and supervision of execution of plans and various projects for the development of the region;
- Review and evaluation of the progress of implementation and to make adjustment in policies and measures as the review may indicate (Sundarbans Affairs Department 2000, 4).

The Board in its formative stage came under the Development and Planning Department (DPD) of West Bengal with the Minister-in-charge of the Sundarbans Area branch of DPD being the chairperson of the directorate. Apart from the minister as chairperson, the vice-chairperson of the West Bengal State Planning Board (WBSPB) and three members of the State Legislative Assembly (MLAs) from the Sundarbans area were ex-officio members of the Board. Although created for the very specific development needs of a region, the Board thrived almost incognito under the overarching administrative structure of the DPD. Except for two short reports, both published in 1979, the Department published no major account of the development activities of the Board. For many years, the Board remained simply an adjunct of the DPD and the situation did not change for the better even after the Left-front came to power in 1977. It took the Board almost twenty-three years to publish its first development report in 1994. And it was only after the Directorate became part of the SAD in 1994 that the report was published. No wonder that in his foreword to the first report S. S. Chattopadhyay, the then Secretary to the DPD, wrote, 'It is customary for a Government department to bring out an Annual Report regarding its functioning once a year. The custom has, however, become almost forgotten. It is, therefore, heartening that Sundarban Development Board has brought out an Annual Report on its own initiative' (Chattopadhyay 1994). In the preface to one of its administrative reports, R. P. Samaddar, former Member Secretary to the Board, described SDB as a 'distinct development agency' (Samaddar 2000) catering to the needs of the deltaic region. Yet, the early history of the Board shows that no sooner was the directorate (Board) formed, than it lost its credibility and initiative as a distinct development agency.

The SAD, of which the Board is now a part, has so far published administration reports covering a period from 1981 to 2005. A recurrent issue, which runs through most of the administration reports, is that the Board since its inception has been plagued by scarcity of funds. The first development fund that came the Board's way was from the International Fund for Agricultural Development (IFAD) sponsored by the World Bank. These funds came around 1981–1982 and continued till 1989. Under the IFAD-assisted project, the activities of the Board were 'restricted' to:

- Creation of sweet [fresh] water reservoirs through re-excavation of silted channels/village ponds, and by closing minor tidal rivers and creeks;
- Improvement of drainage system by constructing a comprehensive drainage system consisting of master sluices, hume pipe sluices, excavation of main drains etc.;
- Improvement of village communication system through construction of brick paved roads, culverts, footbridges, small bridges and jetties;
- Creation of brackish water fish culture, social forestry including mangrove plantation and providing agricultural support service during rabi[4] season (Sundarban Affairs Department 1994, 4).

For the implementation of the IFAD-assisted scheme, the Board was strengthened by the creation of a few divisions such as engineering, social forestry, planning and evaluation, fisheries, finance and administration (Ibid, 3). However, despite the creation of the above divisions, the function of the Board remained one of coordinating the activities of several other specialized departments (such as Public Works Department [PWD], Public Health and Engineering [PHE], Forest and Fisheries) at work in the Sundarbans. One of the two short reports prepared in 1979, much in anticipation of the IFAD funds in 1981–1982, stated that the needs of the Sundarbans are special because it is one of the most backward regions of the state (Sundarban Development Board 1979, 9). The report further stated that the Board set up by an order of the state government is authorized to coordinate all development activities carried out in the Sundarbans by other state government departments (Sundarban Development Board 1979a, 74). It is somewhat surprising that the function of an agency created for dealing with the specific needs of the region is primarily one of coordinating development. Some years later in commenting on the need for such a Board, S. S. Chattopadhyay, the then Secretary to the DPD, in his foreword to the first report of the Board and also of the SAD states:

> Sometimes questions are raised about the raison d'être for a separate agency of development when normal development activities are being carried out by different Government departments in the Sundarban[s] region. Those who ask these questions are not perhaps always aware of the magnitude of the task confronting development agencies in the Sundarban[s] region ... the terrain is very difficult—inhospitable and inaccessible; much of the area is covered by forest infested with the notorious Royal Bengal Tigers; ... the Sundarban Development Board also is a model for harmonious working between the so called bureaucrats and technocrats, for, in this organisation generalists are working side by side with engineers, forest officers, agricultural scientists and statisticians (Chattopadhyay 1994).

The above official depiction tends to show that the problems of the Sundarbans cannot easily be compared with those of other regions of West Bengal. Therefore, a specialized agency in the form of the Board was necessary to address the specific problems of the Sundarbans. Later, in her preface to the third report in 1998, Mira Pande, Secretary to the SAD, reiterates Chattopadhyay's arguments in support of the Board. For Pande, the body is the result of special efforts on the part of the state government to bring about an 'appreciable and consistent improvement in the conditions of the region along with the *normal development activities* which are being carried out by different Government Departments in the region' (Pande 1998; italics added).

Despite the arguments in support of the special significance of the Board, when flipping through the pages of these annual administration reports, one gets the idea that the SDB acts more as a supplementary body than a specialized agency. During its IFAD and post-IFAD days, the Board continued to act as an agency that administered development. In one of his annual budget speeches, the former CPI-M minister heading the Department mentions: '... the Sundarban Development Board was set up for taking up some additional work programmes, in spite of the fact that *normal development activities* of the State Government had been continuing in Sundarbans. The plan proposal for the first year was of Rs. 1 lakh. The Left-front government in their very first budget had allotted Rs. 120 lakh for Sundarban[s] Development' (Mollah 1998, 1; italics added). Thus, the minister, administrators, departmental secretaries are all drawing on the discourses of 'normal development' and 'special problems' of the Sundarbans to argue in support of the Board and its activities. However, these arguments fail to enlighten us about what constitutes the 'special activities' of the Board. Each development report lists the schemes that the Board implements. This includes: (1) providing infrastructure facilities like construction of brick-paved roads, culverts, jetties and bridges, sinking of tube wells; (2) social forestry and tree planting; (3) setting up of small brackish water fish ponds; and (4) agricultural extension programme (mainly rabi and kharif [5] seed distribution to small and marginal farmers). However, in each of the above-mentioned domains, the Board is merely duplicating the efforts and services offered by other government departments such as PWD, PHE (for infrastructure), the Forest Department (for social forestry and plantation), the Department of Fisheries (for brackish water fish and prawn cultivation) and the Department of Agriculture (for agriculture extension programme).

Over the years, there has been a substantial increase in the financial resources of the department. According to the officials of the department almost eighty per cent rise in the budget of the department comes from National Agricultural Bank for Rural Development (NABARD) loan under Rural Infrastructure Development Fund (RIDF). In 2004–2005, the department received additional financial assistance of Rs. 500 lakh from the Government of India (Sundarban Affairs Department 2005) and the 12th Finance Commission of Government of India had given Rs. 100 crores to the department to look into the problems of livelihood in the region. But so far the department has not come up with any comprehensive study on people's livelihood in the region. Despite this increase in financial resources, the Board together with its department still remains a coordinating agency and its development initiatives are far from comprehensive. It is interesting to observe that the department, which collaborates with other government departments on various fronts (agriculture, social forestry, etc.), does not collaborate at all with the irrigation department around an issue as crucial as embankment erosion and displacement. A department constituted for addressing 'specific problems of the region' has failed to come up with a comprehensive rehabilitation policy for the displaced. A question that remains unanswered is why does the Board keep embankment erosion and the human suffering consequent upon it out of its purview? 'We don't do embankments, irrigation is there to take care of it', answered the minister of the SAD, when interviewed. 'But your department does coordinate the activities of many other departments at work in the Sundarbans', I said. 'Yes, we do', he answered, 'but irrigation is a *huge science* [gigantic technological machinery], the department is full of engineers and they are not going to listen to us'. He further stated, 'You must be knowing that people settled in the Sundarbans much before the siltation process was complete. If they had settled later they wouldn't have required embankments. No matter how hard you try, you cannot conquer nature. Once you claimed land from nature and now nature is claiming land from you'.

Under these circumstances, the department's agriculture extension programme, which consists largely of providing rabi and kharif seeds to farmers and popularizing cotton and mushroom cultivation among the farming households, remains far from being a success. The question of sustainable agriculture is deeply connected to the ecology of the delta. With people continuing to lose lands and being forced to live on the edge of embankments, agriculture remains of marginal significance. The success of mushroom as a second crop in winter is also very limited. Women living in remote islands find it difficult to market their products even when they are trained in mushroom farming. Despite the minister's repeated emphasis in the departmental budget speeches on mushroom's protein content (Mollah 1993, 1994) the crop is yet to become a part of the diet of the islanders. In the absence of any demand for mushrooms in the region, its prospect as a second crop is unlikely to succeed in the long run. For the plan period 1996–2000, the Tenth Finance Commission had recommended a grant of Rs. 35 crores for the *overall development of the Sundarbans*, and the SAD had been authorized to act as a *nodal department* for the said grant (Sundarban Affairs Department 2000, 3; italics added). Subsequently, the SDB had prepared an action plan for the utilization of this grant. As a nodal agency, it had clearly specified the departments that it would collaborate with. The only department not mentioned in this report was Irrigation and Waterways, which seems to suggest that the SAD can continue to remain a nodal agency and attempt to do what it considers to be comprehensive development of the region without having to address an issue as endemic as embankment erosion and flooding. I turn now to a consideration of the ways in which irrigation department addresses these problems.

The irrigation department and embankments in the Sundarbans: Discourses of erosion

In the month of November 1988, a cyclone hit the islands of the Sundarbans. It was a severe storm with a wind speed of 250 kilometres per hour. In March the following year, while presenting the annual budget in the West Bengal Legislative Assembly, the Minister of the Irrigation and Waterways Department reflected on this incident:

> [Honourable] Members of the house are aware that a severe cyclone hit West Bengal coastal areas on 28th November 1988 ... As a result, there has been extensive damage to flood protecting embankments and structures in the estuarine areas of South 24-Parganas and parts of North 24-Parganas and Midnapore district due to wave action.
>
> The extent of damage to embankment sluices and structures under my Department has been estimated at Rs. 604.00 lakhs. Repair works at vulnerable areas have been completed with available fund amounting to Rs. 55 lakhs. The residual portions of the repair and restoration works are in progress with target of completion by March 1989 (Bandyopadhyay 1989, 12–13).

The minister's words were hardly indicative of what the islanders actually experienced on the day of the cyclone. The minister estimated the damage the cyclone caused to his department's property such as sluices and embankments, but there was no mention of how many people actually died. The cyclone had occurred during the harvesting season when the plants were fully grown. After the cyclone, devastated rice fields and rivers filled with corpses of human beings and livestock were what the survivors of the Sundarbans were left with.

Soon after this disaster the newsmagazine *The Week* in its edition dated February 1989, carried a report on the cyclone and its effects on life in the Sundarbans:

> Last year's cyclone of November 29, blowing at 250 km an hour ... was one of the fiercest that hit southern Bengal in recent times. The cyclone, which killed 500 people and damaged crops worth Rs 83 crore, was the harshest in South 24-Parganas district where 400 people perished and the Gosaba block

was ravaged ... Half a dozen state government ministers had made a whirlwind tour of the region devastated by the cyclone ... but the villagers got nothing out of these exercises (*The Week* 1989, 36–39).

Starvation and impoverishment continued after the cyclone because apathy and neglect permeate development initiatives in the Sundarbans. This apathy informs both the policies and practices of the irrigation department.

Fifty-four inhabited islands in the Sundarbans have 3,500 kilometres long earthen embankments surrounding and protecting them. Since 1960, soon after the abolition of the zamindari system, the protection and maintenance of the Sundarbans embankments has been the responsibility of the Department of Irrigation and Waterways, which is under the charge of two ministers: a fully fledged cabinet minister heading this department and a minister of state for irrigation. The Department of Irrigation and Waterways has an elaborate bureaucratic structure. A Secretary, who occupies a position subordinate to that of ministers and is responsible for the running of the entire directorate, is the administrative head of the department. Then the whole directorate is divided into a number of functional zones or circles. However, the show is actually run by a host of civil and mechanical engineers who are the backbone of the department. Embankments in the Sundarbans and their maintenance come under the Eastern Circle with a Superintendent Engineer in charge of each circle. A circle is further divided into a number of divisions. An Executive Engineer is in charge of each division. Kusumpur and Gosaba in the Sundarbans come under Joynagar Division. A division is then divided into a few sub-divisions with a Sub-divisional Engineer responsible for each sub-division. Finally, a sub-division is divided into a number of sections and a Section Officer is appointed for each section.

Embankments erode mainly due either to wave pressure or river currents causing breaches at the bed of the river. The protection of these embankments is financed by the 'flood control' sector of the departmental budget and under 'flood control' sector comes the 'urgent development work' in the Sundarbans. On paper 'urgent development work' includes the following:

- Strengthening the embankment against wave pressure;
- Laying bricks on the slopes of embankments and anti-erosion work;
- Constructing new sluice gates and renovating existing ones to improve drainage conditions;
- Building ring embankments[6] (Irrigation and Waterways n.d.; translated from vernacular).

Embankments spell disaster for the people, since breaches occur all the time. Yet, as will be seen, what actually happens in the name of urgent development work is either indiscriminate land acquisition on the pretext of building ring embankments or hastily done patchwork repair of breaches that is likely to lead to problems in future.

Although the irrigation department does anti-erosion work in the Sundarbans, in the department's discourses of development some kinds of erosion are seen as more menacing than others. In the departmental budget speeches (1989–2013), concerns had been expressed over the growing Ganga–Padma[7] erosion in Malda and Murshidabad[8]. A considerable section of the departmental budget proposal is devoted to an analysis of this problem. The minister considers this problem menacing because it not only engulfs thickly populated villages and results in the loss of fertile agricultural land, but also endangers national property like railway tracks, National Highway 34, the Feeder Canal at Farakka Barrage and many places of archaeological, historical and religious importance (Bandyopadhyay 1989, 1990, 1994). Therefore, an action plan for the prevention of erosion estimated at Rs. 355 crores was presented for consideration of the Ministry of Water Resources, Government of India and the Planning Commission in 1992 (Bandyopadhyay 1993). In 1998–1999, 1999–2000 and 2000–2001, Rs. 60 crores, Rs. 22.50 crores and Rs. 8 crores respectively were earmarked to tackle the Ganga–Padma erosion (Bandyopadhyay 1998, 1999, 2000). Because the Ganga–Padma erosion is destroying national property, the minister suggests that this erosion be given national importance. However, the same department remains silent over the problem of erosion in the Sundarbans[9]. Even after the disastrous cyclone of 1988, the department did not bother to devote even a page of the budget proposal to the problem of land erosion in the Sundarbans. Because erosion here affects only the property of local people and not of the 'nation', it does not assume national importance.

Closely linked to these differing notions of erosion is also the question of land acquisition. Every year in the annual budget proposal, the Minister-in-charge heading this department appeals to the state government to generate funds for the development and maintenance of big dams like the Teesta, Mahananda, Subarnarekha and Damodar Valley Projects. These dams, according to the minister, are of national importance and therefore he suggests that adequate attention be given to them. Regarding the Teesta Barrage Project in north Bengal, the minister repeatedly emphasizes the need for land acquisition with suitable compensation (Bandyopadhyay 2000, 8) to facilitate the speedy completion of the project. In this connection, he requests the panchayat and land reforms departments to cooperate with his department. A joint endeavour of these two departments, the minister suggests, would aim to shorten the time-consuming land-acquisition process and create a sense of identification of people with the projects (Ibid).

In the Sundarbans, land acquisition also constitutes an integral aspect of embankment construction and maintenance. In the event of embankment collapse, the irrigation department acquires land for constructing a ring embankment. But in most cases, the department unilaterally acquires land for the purpose of such embankment construction or even repair work. The irrigation department does not find it necessary to collaborate with the panchayat and land reform departments to sort out the complex issues of land acquisition, compensation and rehabilitation. Here, land is acquired on the pretext that such acquisition is necessary to provide a greater protection to the population.

Building embankments: A context for governmental action, acquisition and apathy

Embankment maintenance can also be seen, to use Wade's phrase, as 'a specific context of governmental action in the countryside' (Wade 1982, 287). To understand government action, I will concentrate on Kusumpur and its neighbouring islands. As stated earlier in this chapter, the northern part of Kusumpur village was erosion prone with the rivers eroding banks on both sides. Families living in this narrow stretch had survived four ring embankments. Each time a ring embankment is constructed, it encroaches upon people's lands, ponds and even houses. When a considerable stretch of embankment collapses, the obvious solution lies in building a ring embankment, which is normally built behind the old one that has collapsed. Construction of such ring embankments provides the engineers and contractors with an opportunity to make money. Since land acquisition is necessary for building new embankments, decisions about how much land would be acquired or how far behind the existing embankment the new one would be built are left entirely to the discretion of the engineers. They justify such acquisition on the pretext that it is they who are better able to judge what is 'good' for the people. With the help of contractors, the engineers acquire land for building such ring embankments. Many people in north Kusumpur had lost their lands and houses due to multiple ring embankments but none of them had received any compensation for their loss. 'One more ring embankment and we are all gone', said Haren Mondal, a resident of north Kusumpur. 'These engineers think no end of themselves. Most of the time they remain in Kolkata to spend time with their families and are never available in their Gosaba office', Haren further stated.

On the eastern side of Kusumpur right across the river Goira was Garantala island of Gosaba block. In 1999, the embankment collapsed in Garantala island, which was under water for a month. The irrigation department issued an order for building a 1,500-feet wide ring embankment that rendered about ninety families landless and about thirty of them homeless (see Map 5). Soon after the embankment collapsed, the villagers themselves decided to build a temporary protective embankment, which they said, would have required a much smaller area of land. However, when the villagers proceeded to rebuild that embankment, the irrigation department engineers and local panchayat leaders intervened, and the villagers were conveniently sidelined. 'It's a pact between the contractors and engineers', said Bharat Mondal, an aggrieved resident of Garantala, who lost all his land because of the new embankment. 'The contractor employed about one hundred labourers. Many of them haven't got their payments as yet, but the contractor has left with all his dues. The irrigation department has its own list of contractors. They are selected on the basis of whether they are ready to give the officers a share of their profit. We pleaded with the engineers that such a huge ring was not required, but they refused to listen to us. All the officers in the irrigation office had taken money from the contractors'. Bharat's narrative revealed people's suspicions about the intentions of the engineers and contractors. The engineers decide to build the new ring embankment as far behind the existing one as suits them. People's lands, ponds and even houses are annexed for this purpose. As I have said earlier, no compensation is ever paid to them.

The irrigation department operates through its many field offices in the Sundarbans, one of which is on Gosaba island. The engineers in charge are meant to provide emergency services in the event of a disaster, but when disaster strikes they are rarely to be found in their offices. According to the gossip doing rounds at the tea stalls of Gosaba, the engineers mostly remain on 'sick leave' and usually only a caretaker is there to look after the office. In the event of a disaster, people's suffering is heightened when they come all the way from a distant island of Gosaba block only to find that the engineers who are supposed to help them are away in Kolkata.

After several failed visits, I finally managed to get hold of an engineer of the Gosaba irrigation office. When I reached the office it was 2:30 in the afternoon.

The office was practically empty and I found the same person, whom I had encountered on my earlier visits to the place, sitting inside and dozing. He was from Kusumpur and worked in this office on a temporary basis. He woke up when I knocked on the door. Seeing me he smiled and before I could say anything he said, 'You must be looking for saheb'. Pointing in the direction of a house next to the office he said, 'He is in his house'. I found Dinu, a dafadar[10] from north Kusumpur, sitting at the doorstep of the section officer's house. Seeing me he gave an embarrassed smile. 'Sir [officer] is inside, should I go and call him?' Pointing to the bottle in his hand, I asked him, 'What is that?' 'I came to give Sir the honey he asked for', he answered. 'It is difficult to get honey at this time, but I managed some'.

Meanwhile the engineer came out and asked me to take a seat on the small wooden bed placed against the wall of the balcony, which served as an entrance to the inner part of the house. The bed was covered in files and papers. The engineer cleared them to make some space for both of us. It was easy to see that he rarely visited his office and had effectively turned the balcony into his office. I sat on the cot with my back to Dinu who squatted on the steps outside. Seeing Dinu with the bottle the engineer said, 'That's very little, get some more', and then turning to me said, 'My son loves honey. That is one reason why I wish I could stay here'. 'Why, are you leaving the Sundarbans?' I asked. 'Yes, after all this delay I finally got my transfer order', he answered.

When I asked him about the Garantala embankment he was visibly annoyed. 'See, this is the problem with these people. They do not appreciate what you do for them. It's because of me that today it is only 1,500 feet, otherwise it would have been much more'. He then turned to Dinu and asked, 'Is it true or not?' and without waiting for his confirmation continued, 'I persuaded the divisional engineers and overseers not to encroach upon their land. These people think that they would have constructed a better embankment. If the villagers can do everything why does the government need engineers and an irrigation office?' Turning again to Dinu he asked, 'Is what I have said true or not? Anyway I have had enough with the Sundarbans. I will be leaving soon. I was here for five years and let me tell you if you are here for more than two years you will end up having severe heart problems. Work here involves so much tension and anxiety ... the less said the better', the engineer shook his head in utter disgruntlement.

During the conversation Dinu was being used as a sounding board. The engineer knew exactly what his responses would be. Dafadars are recruited for supervising repair or construction work. Since local irrigation engineers are the key factors in the appointment of dafadars, they have to be assured of a constant flow of goods and services. It is the duty of a dafadar to keep his superiors happy, lest he run the risk of losing his job. Therefore, Dinu's periodic visits with a bottle of honey and his preparedness to be at the beck and call of the engineers are indicative of the way state bureaucracy operates at the local level and also of the effects of the state on the everyday lives of the rural people.

While I was in Kusumpur, the divisional engineer of Joynagar Division[11] happened to pay a visit. I got the news while I was talking to a few workers near an embankment site in Kusumpur. 'How often do they visit?' I asked. 'Sometimes not even once a year', answered Monoranjan, who was busy repairing the slopes of the embankment. I further asked, 'Why do you think he is coming?' 'God knows why he is coming', Monoranjan said, 'come tomorrow and see for yourself'.

Early next morning, I went to north Kusumpur. The monsoon had already set in and it was raining quite heavily. The roads and embankments were extremely slippery and muddy. Villagers there had set up a makeshift tent on the embankment and were all waiting inside. I went into the tent and waited for the engineer, who was supposed to come at ten o'clock in the morning. It was a long wait and even by around two o'clock in the afternoon there was no sign of him. I skipped lunch thinking that if I went for my lunch I would miss him. People around me became restless and many went home. Around five o'clock in the evening, we suddenly saw a launch approaching the jetty at Kusumpur. As the launch stopped at the jetty, the section officer got off and came up to the crowd and said, 'Saheb has come all the way from Kolkata and is too tired to meet you all now. Tonight he will stay at the irrigation guest house at Gosaba and will come tomorrow morning for a visit'. The engineer was sitting on the launch deck and was happily chatting away with his colleagues. Next morning the launch came again and it was raining even more heavily than the day before. The engineer and his two subordinate staff (overseer and estimator) got off the launch but could not proceed further. They could not stand upright on the muddy slopes, let alone walk through the terrain. Ultimately, they had to be lifted by the villagers and carried to the spots the villagers wanted them to visit. From their faces, one could make out that they were in no mood to visit the site and desperately wanted to get back to the launch. The much-awaited inspection was over in five minutes. They told the panchayat secretary, 'We have seen the spots, they are bad. We need to get back to the Gosaba office to discuss it amongst ourselves'.

The above incident points to a lack of concern on the part of the so-called developers who do development in circumstances with which they are not familiar and avoid situations that have the slightest potential for questioning their wisdom. This reminds us of Hobart's thesis: 'The idea of "underdevelopment" itself and the means to alleviate the perceived problem are formulated in the dominant powers' account of how the world is. The relationship of the developers and those to-be-developed is constituted by the developers' knowledge and categories, be it the nation-state, the market or the institutions which are designed to give a semblance of control over these confections' (Hobart 1993, 2). However, while considering the activities of developers as a category we should also keep in mind the practices of those officials or functionaries who are more localized than their seniors. For example, the section officers or the sub-divisional officers based at the Gosaba irrigation office, who accompanied their superior officers during their visit, walked through the muddy terrain with their trousers rolled up to their knees. While their superiors were lifted and carried to the spots for a visit, the field officers seemed well conversant with the muddy terrain. Unlike the executive engineers and overseers of the division, they did not remain uninformed about the area they administered. On the contrary, their familiarity with the local conditions worked to their advantage enabling them to maintain contact with the local panchayat and party leaders, to strike deals at various levels in the local political circle and have locals like Dinu at their beck and call.

Cyclone Aila: The politics of relief and aid

In 2009, another cyclone struck the Sundarbans. Basanti Raptan, a resident of the southern part of Kusumpur island in Gosaba block, woke up to a morning that was different but not unusual in the Sundarbans. Since early morning, a thick cloud hung over the island and there was a strong wind blowing across the river. Basanti woke up early to start her household chores little realizing what the day had in store for her. She lived with her one-and-half year old daughter as her husband was away in Kolkata to work as a construction labourer. While Basanti got up, her daughter was still asleep. As the day progressed, the wind began to blow harder. Around eleven in the morning the wind suddenly changed into a violent storm. The mud wall and doors of her house started to tremble due to the impact of the storm, and Basanti could see from her courtyard that the storm made coconut trees bend into halves. Suddenly, she heard cries from her neighbours' houses. Before she could step out, a huge wave of water broke open the door and mud wall of her courtyard and pushed her into a corner. In complete bewilderment, Basanti clutched on to the mud wall of her kitchen. Seeing the water rising menacingly, she waded through waist deep water in a desperate bid to rescue her daughter who was still sleeping. Before she could reach her room, a fresh wave of water broke down the mud walls of the room and swept the child away from her. Basanti's cry for help was lost in the deafening sound of wind and water.

I heard this incident during my conversation with the villagers at a tea shop in Kusumpur. On 25 May, 2009 twenty years after the cyclone of 1988, another devastating cyclone, Aila, swept across south Bengal, particularly the deltaic Sundarbans, killing people and their livestock and rendering thousands homeless. Those living on the margins once again became marginalized. The very next day islanders were found lined up on the embankment pleading, shouting and jostling each other trying to grab relief and aid that came their way. And many others having lost their land, houses and also their family members already started to migrate out of the Sundarbans in search of an uncertain future in Kolkata. The saline water that broke through embankments flooded the villages, destroyed mud houses and polluted rice fields.

> Hamlets have been reduced to wasteland – with submerged crops, uprooted trees, shattered homesteads and emaciated [and dead] cattle all around. Ponds which have been the only source of portable water lay contaminated and stinking. Not even stray dogs that survived the disaster would go near them (Chattopadhyay 2009, 33).

In South 24 Parganas, the Sundarbans blocks that were badly hit by Aila were Gosaba, Basanti, Sagar, Namkhana, Patharpratima and Kakdwip. The total length of embankments severely damaged in these blocks was 621.95 kilometres (Irrigation and Waterways Department n.d.a). Out of 308 sluice gates, about 125 were completely damaged resulting in saline ingress and flooding of the islands (Ibid). The total area inundated with saline water in South 24 Parganas was 105,075 hectares (Department of Agriculture and Horticulture of South 24 Parganas n.d.). The devastation of Aila put the government on red alert. As part of its short-term relief, the government distributed food, clothes, water and medicines among the Aila victims. Local sub-divisional, block development offices and panchayat bodies were mobilized for immediate relief activities. Government initiatives were supplemented by local, regional and even international NGOs.

The cyclone occurred at a time when the Left-front's thirty-four years of rule in West Bengal showed signs of erosion. The front lost majority of its parliamentary seats to the Trinamul Congress, the main opposition party which had already made significant inroads into different districts of West Bengal. Having been successful in winning all the major left strongholds in the Sundarbans, Mamata Banerjee, the TMC leader and the then Railway Minister of the Congress-coalition government at the centre wanted to strike her party's roots further into the Sundarbans soil. She was active in organizing relief for the victims and also wanted to convey the image of the Left-front government as being incapable of responding to the needs of the poor and vulnerable. To this end, she even requested the central government not to send the Aila relief fund through the state government and instead channel the fund through the South 24 Parganas district panchayat body (i.e., Zilla Parishad)[12] where her party was in power.

At this point, it is interesting to see how the Left–Trinamul rivalry articulated through governmental aid and relief for the Aila victims. Instead of viewing political parties as discrete entities at work in liberal democracies, I see them as important agents implicated in the ways in which governmental power is exercised and rule is consolidated. To understand how Aila became a site for the pursuit of divergent and conflicting developmental interests, I focus on regional and local newspaper reports. Ever since Aila struck the Sundarbans, Kolkata-based newspapers such as *The Telegraph*, *The Times of India* and *Anandabazar* had been publishing reports on Aila and the politics of aid and relief. *The Telegraph* dated 7 July, 2009 carried a report which stated:

> [The central Finance Minister] Pranab Mukherjee today proposed to allocate Rs 1,000 crore for Aila relief to the Bengal government as sought by Chief Minister Buddhadeb Bhattacharjee, a move likely to please the state's Left-Front regime more than ally Mamata Banerjee ... Bhattacharjee had requested the Prime Minister to release Rs 1,000 crore from the national calamity and contingency fund for reconstruction and relief in areas hit by the cyclone on May 25. But Mamata and her party [Trinamul Congress] had demanded that central funds be disbursed directly to panchayat officials bypassing the state government. The Trinamul chief had said that any money provided to the CPM-led state government would only reach party supporters (*The Telegraph* 2009a).

The very next day *The Telegraph* published another report highlighting left–Trinamul rivalry over Aila aid:

> The Centre has sanctioned Rs 478 crore from its national calamity fund for Aila relief over and above the money [Rs 1,000 crore] promised in yesterday's budget.

State finance minister Asim Dasgupta said [that] the Rs 1,000 crore allocation ... was meant for construction of concrete embankments and the money sanctioned today for immediate repair and rebuilding jobs ... Mamata Banerjee had pleaded with the Centre not to channel the aid through the state government. She wanted the money to go directly to the panchayats – "PM to DM" – many of which are in Trinamul Congress control now (*The Telegraph* 2009b).

A news report published in *The Times of India* dated 30 October, 2009 gave a further twist to partisan interests crystallized over Aila aid:

Trinamool Congress zilla parishad sabhadhipatis of South 24 Parganas and East Midnapore skipped the meeting with the Chief Minister on Wednesday, ... They wanted to "prove" that the Rs 100 crore allocated by Prime Minister Manmohan Singh from the PM's National Relief Funds for Aila victims was due to Mamata Banerjee's initiative. Trinamool leader Sobhan Chatterjee displayed a letter written by the PM to Mamata Banerjee about funds for cyclone shelters. He accused CPM of playing politics with rural development ... (*The Times of India* 2009).

As we turn our attention away from regional newspapers to local newspapers published from the Sundarbans we see a different scenario, one replete with party rivalries amidst which it is the victims who suffer. *Badweep Barta* dated 1–30 June, 2009 provided an account of people's condition in the Aila-struck Sundarbans and stated that because of partisan and vested interests, governmental relief had not reached the victims living in the remote parts of the Sundarbans islands. Relief and money are being channelled through local government offices or panchayat bodies, but effectively it is the party leaders (of the CPI-M, RSP or TMC) who control people's access to aid and relief. In sheer desperation, people are leaving the Sundarbans in search of food and shelter elsewhere (*Badweep Barta* 2009a).

Badweep Barta in its news report dated 16–31 August, 2009 further highlighted politics over pond desalinization. It stated:

People of Masjidbati mouja in Basanti are devastated by Aila. The local Trinamul leadership has accused [the RSP] panchayat of playing dirty politics over Aila relief. As a result people who are really in trouble have not got any relief. Local Trinamul leader Mannan Sheikh alleged that the money that came for desalinizing the ponds has been improperly used to desalinate only 40–50 ponds. It is further alleged that a list of beneficiaries has been prepared for governmental compensation. As per rule the enlisted people are required to open bank accounts. The panchayat has made the Aila victims pay Rs. 100

each for opening a bank account. Without this money compensation will not reach the victims (*Badweep Barta* 2009b; translated from vernacular).

Even though the TMC and its ally the SUCI had been successful in winning parliamentary constituencies (Joynagar and Mathurapur) of the Sundarbans, Gosaba and Basanti still remained left (particularly the RSP) strongholds (see Tables 3.1–3.4 for the relative strength of the left parties and the TMC). The Assembly constituency-wise breakup of Parliamentary constituency of Joynagar shows dominance of the RSP in Basanti and Canning East and a close contest between the left and the TMC-supported SUCI in Gosaba. At the panchayat level, the Zilla Parishad was in TMC's control, but the block and village panchayat scenario in Basanti and Gosaba explains why there was intense power tussle between the RSP and TMC. The local level conflicts and rivalries over Aila relief can be seen as expressions of the TMC making inroads into the left bastions. This gets reflected in a local newspaper *Aranyadoot*, which stated that the TMC was gradually increasing its strength in Basanti block. About 400 CPI-M and RSP cadres had joined the TMC in this block. This floor crossing was clearly a slow but steady process (*Aranyadoot* 2009a). Similar processes were at work in Gosaba block also where many RSP and CPI-M cadres were known to have joined the TMC (*Aranyadoot* 2009b). It is perhaps needless to say that this floor crossing had intensified party rivalries and confrontations over Aila relief.

Table 3.1: South 24 Parganas' Joynagar (Reserved) Parliamentary Constituency's Assembly-wise breakup of the election results in the Sundarbans.

Name of the Assembly Constituency	Nimai Barman – RSP Candidate	Tarun Mondal – TMC-Supported SUCI Candidate
Gosaba (Reserved)	60,908 votes	64,403 votes
Basanti (Reserved)	64,434 votes	46,397 votes
Kultali (Reserved)	64,993 votes	70,116 votes
Joynagar (Reserved)	36,114 votes	73,043 votes
Canning West (Reserved)	45,600 votes	73,861 votes
Canning East	71,358 votes	50,114 votes
Magrahat East (Reserved)	49,040 votes	68,239 votes
Total votes	392,447 votes	446,173 votes
Postal ballot	48 votes	27 votes
Total votes polled	392,495 votes	446,200 votes

Source: *Aranyadoot* 1–14 August, 2009c.

Table 3.2: South 24 Parganas' Mathurapur (Reserved) Parliamentary Constituency's Assembly-wise breakup of the election results in the Sundarbans

Name of the Assembly Constituency	Animesh Naskar – CPI-M Candidate	C. M. Jatua – TMC Candidate
Patharpratima	71,611 votes	87,305 votes
Kakdwip	63,570 votes	78,089 votes
Sagar	76,951 votes	88,862 votes
Kulpi	51,458 votes	71,272 votes
Raidighi	69,401 votes	88,439 votes
Mandirbazar (Reserved)	55,064 votes	81,685 votes
Magrahat West	47,133 votes	69,421 votes
Total votes	435,188 votes	565,073 votes
Postal ballot	354 votes	432 votes
Total votes polled	435,542 votes	565,505 votes

Source: *Aranyadoot* 1–14 August, 2009c.

Table 3.3: Electoral strength of parties at the village panchayats in Gosaba and Basanti blocks

Block	No. of Village Panchayats	CPI-M	RSP	TMC	Congress
Basanti	13	3	9	1	-
Gosaba	14	2	6	6	-

Source: Gosaba and Basanti Block Panchayat offices.

Table 3.4: Electoral strength of parties at the block-level panchayat (Panchayat Samity)

Block	No. of Members	CPI-M	RSP	TMC	Congress
Basanti	38	10	24	3	1
Gosaba	37	6	21	10	-

Source: Gosaba and Basanti Block Panchayat offices.

Irrigation department and Aila struck embankments: Apathy continues

While addressing the members of the West Bengal Legislative Assembly on 16 June, 2009 the RSP minister heading the irrigation department of the

Left-front government reflected on the cyclone Aila and its devastating impact on the Sundarbans:

> I want to draw attention of the honourable members of the House to cyclone Aila that struck the Sundarbans … The cyclonic depression in the Bay of Bengal caused the water body to rise to unprecedented heights, destroying the mud embankments and resulting in flooding of the inhabited islands … The retired chief engineer of my department who was also the secretary of the irrigation department is holding talks with the representatives of the World Bank to find a solution to the problem … Meanwhile my department has started repairing embankments in many places to prevent further saline ingress during high tides. However, land acquisition at times poses as a major obstacle to the process of embankment construction and repair (Naskar 2009; translated from vernacular).

Central Water Commission, Government of India, in collaboration with the Irrigation Department, Government of West Bengal, had constituted a Task Force on 11 June, 2009 to suggest short-term measures to rehabilitate the damaged and washed-away embankments and long-term measures to prevent occurrence of embankment failure in future (Ministry of Water Resources, Government of India 2009, i). The Aila Task Force further stated:

> … due to cyclone 'AILA', about 1000 km length of existing embankments have been affected for which the short term measures have been advised by the Task Force. But for long-term measures, the status of remaining 2500 km long embankments are to be reviewed, in order to have continuous stable embankments along the river banks or sea shore and therefore, the design and construction methodology for remaining embankments may be worked out … requirements of funds in respect of restoration works as "Short Term Measures" are already projected by the Ministry of Water Resources to the Planning Commission for enhancement of plan outlay to Rs. 8,000 crore, under "FMP" [Flood Management Programme] scheme. For additional funds required for long-term measures, International Financial Institutions may have to be approached (Ibid, 12–13).

The irrigation minister's statement in the West Bengal Legislative Assembly together with the recommendations of the Task Force hinted at the possibility of further land acquisition for the purpose of building or rebuilding the Aila-damaged embankments. Embankment rebuilding is of crucial importance in post-Aila Sundarbans. And there is no denying the fact that land acquisition is necessary to facilitate embankment rebuilding.

Earlier in this chapter, I discussed the actual practice of land acquisition and how such acquisition works to the advantage of the irrigation engineers and contractors and thereby reduces people to mere objects of governmental power. The pre-Aila northern part of Kusumpur island was already erosion prone. Families living in this narrow stretch had survived four successive ring embankments. The gigantic governmental department with its elaborate blueprint for land acquisition encroached upon people's lands, ponds and even houses. No compensation ever came their way. Before Aila, the maximum land available between the rivers on both sides was 800 feet (this fact is mentioned at the beginning of the first chapter). Aila had further narrowed the width of this land. About 130 families living in this narrow stretch of land were almost on the verge of being displaced.

I visited Kusumpur after Aila. The families I met in north Kusumpur looked anxious and pensive. Most of the families had lost their houses and lived in temporary tents they had set up near their broken huts. Their rice fields were still waterlogged. Saline ingress had contaminated their ponds. When I asked them about rebuilding of embankments and the possibility of land acquisition, they had mixed responses. Although they accepted that government would need land to rebuild the damaged embankments, they also made it clear that they needed compensation for their relocation. The possibility of further land acquisition made them feel apprehensive about their future in north Kusumpur. They concluded that a ring embankment would split Kusumpur into two separate islands. And this meant that residents of north Kusumpur would have to leave their habitat. Islanders' past experiences of land acquisition had made them suspicious about the government's intentions. Aila had already deprived them of the means of livelihood. Without much money left in their coffers, these people were on the verge of facing an existential crisis.

Ever since Aila occurred in May 2009, and the Aila Task Force was constituted to look into the problems of the islanders, not much had happened in the Sundarbans. Even the short-term embankment building measures recommended by the Task Force were not addressed in the region. In 2011, the Left-front was thrown out of power by the contending TMC with Mamata Banerjee as the new Chief Minister of West Bengal. But development apathy continues in the Indian Sundarbans. *The Telegraph* on 19 July, 2013 carried a report which stated:

> The state irrigation department was supposed to build the embankments but the plan has remained largely on paper ... Soon after the Aila struck in May 2009, the central government formed a task force to ... take measures

to secure the coastal region from the onslaught of cyclones ... "After the report was submitted, the government decided to build embankments along 740 km of the Sunderbans to save the villages on the coast. A Rs 5,030-crore project was sanctioned. Barely 15 per cent of that work has been done." (*The Telegraph* 2013).

In 2009, the Left-front government seemed to have worked out a compensation package for 6,000 acres to be acquired in the Sundarbans to build embankments. There was no official announcement on the amount to be offered but it was learnt that the amount could be between Rs. 5 lakh and Rs. 6 lakh an acre. According to the package, landowners would get the value of the land fixed by the government plus thirty per cent of it as solatium and an interest of twelve per cent a year from the date of declaration of valuation till the disbursal of cheques. Registered sharecroppers would be offered fifty per cent of the value of their land and Rs. 34,000 (*The Telegraph* 2009c). The compensation proposal that the Left-front government contemplated in 2009 did not include the landless agricultural labourers who were not registered as sharecroppers. When one looks at the plight of the villagers in north Kusumpur in the light of the compensation proposal, one feels that many families would not stand to gain from this proposed compensation package as land acquisitions in the past already robbed them of their land and livelihood. Therefore, a question arises as to what happens to them in the event of their being forced to leave their habitat. However, what is more significant to note is that this proposed package was never actualized by the left government and ever since the TMC came to power nothing comprehensive was done on the embankment front in the Sundarbans.[13]

Development as protocol

The Task Force report on the restoration of embankments in post-Aila Sundarbans was an important document prepared by the Ministry of Water Resources, Government of India in collaboration with the Irrigation and Waterways Department, Government of West Bengal. The report of the Task Force evolved out of meetings held at the irrigation department in Kolkata. These meetings were chaired by the former Chairman, Central Water Commission and attended by irrigation engineers, technology experts from reputed institutions in India, NGO functionaries and so on. These experts visited Aila-struck areas of the Sundarbans and inspected the damaged embankments. The annexure of the report was filled with

the minutes of each meeting held in Kolkata. The meetings usually started
with thanks to the Governor of West Bengal and the Finance Minister of
the Government of West Bengal for their guidance and support and ended
with vote of thanks to the chair. The minutes indicated that the experts
deliberated upon various aspects of the Sundarbans embankments. The
meetings prepared costs involved in restoring damaged embankments.
Apart from indicating cost to be incurred for dredging rivers and restoring
embankments and sluices in various places in the Sundarbans, the report
also mentioned a section on miscellaneous and overhead costs (Ministry
of Water Resources, Government of India 2009, 28). Miscellaneous and
overhead costs refer to the maintenance of the infrastructure of the irrigation
department. Such costs include expenditure the department incurs for
setting up its offices at the restoration sites, recruiting personnel (such as
ground officials, caretaker, security staff) for maintaining those offices,
disbursing salaries of the recruited officials and so on. The estimated
miscellaneous and overhead cost mentioned in the report was Rs. 346.11
crores. Thus, even before the irrigation department needed money to
undertake restoration of the Sundarbans embankments, the department
needed the above amount to sustain its gigantic machinery.

Looking at the Task Force report, one is reminded of Ferguson's *The
Anti-politics Machine* (1990) in which he explores the idea of development
in connection with his analysis of a rural development project in Lesotho
in Africa. Ferguson shows that, despite the failure of the project funded
by the World Bank, development remains central in offering insights into
the ways social relations are structured and power is deployed by the state
and the international aid agencies. Development has achieved the status
of a certainty in the social imaginary (Escobar 1995, 5). Development
remains central to constructing underdevelopment. Development remains
central in explaining why a sense of apathy persists and informs policies
towards people's welfare in the Sundarbans. If one task force committee
fails to deliver, another task force committee is set up to look into the
problems of its predecessor. Development remains the pivot around which
revolve disaster, land acquisition, displacement and deprivation in the
Sundarbans. Interestingly, development also becomes an occasion for
understanding the activities of the islanders, the victims of disaster and
development, who work around the governmental machinery. Practices
of embankment workers, dafadars, beldars and prawn seed collectors
crystallize only in the context of governmental activities related to

embankment maintenance, construction and flood control. In a land where livelihood options are limited, the embankment, which is a source of disaster for the islanders, also provides them with opportunities of livelihood. Thus, it is the prospect of livelihood, even in times of crisis and vulnerability, which makes possible transactions and negotiations between these categories of local people and the governmental machinery. The subsequent chapters focus upon livelihood strategies and practices of these people.

Notes

[1] *Mangal* refers to mangrove. Since the seminar mainly revolved around the Sundarbans ecosystem and mangrove vegetation, the volume was entitled *Sundarbans Mangal*.

[2] Expenditures, spending and values are mostly shown in Indian National Rupee (INR) unless specified otherwise. Currently a dollar is equal to INR 67.

[3] The word Sundarbans is used in plural throughout the book. However, the use of the word 'Sundarban' in Sundarban Development Board and Department of Sundarban Affairs is used in singular.

[4] Rabi refers to winter crops.

[5] Kharif refers to crop grown during monsoon.

[6] When an old breached or collapsed embankment is replaced with a new stretch of embankment, it is normally built behind the old stretch inside the village. Because the new stretch is crescent shaped, it is called ring embankment.

[7] The Ganga is a major river of northern India. It flows through Uttar Pradesh , Bihar and finally into West Bengal. Near Farakka in Murshidabad in West Bengal, the Ganga divides and flows in two directions: one into Bangladesh (called the Padma) and the other towards the south of Bengal (called the Hooghly).

[8] Malda and Murshidabad are two districts of West Bengal, both being situated on the banks of the Ganga.

[9] The irrigation department's focus on Ganga-Padma erosion and big dams has remained unaltered in the subsequent budget speeches of the ministers of the department (see Naskar 2006, 2007, 2008, 2010). Even with the coming of the Trinamul Congress led government the department's focus on big dams remains unaltered (see Bhunia 2011, 2012).

[10] Dafadars are temporary labour supervisors recruited by the field offices of the irrigation department in the Sundarbans. Such recruitments are made from the local villages. But once a person is recruited as a dafadar, he continues in the post.

[11] Out of 3,500 kilometres long embankments in the Sundarbans, Gosaba block accounts for 372.5 kilometres. Joynagar division is responsible for the protection of this particular stretch of the embankment.

[12] Zilla Parishad is the district level panchayat body (uppermost tier of the three-tier panchayat structure in West Bengal). The South 24 Parganas Zilla Parishad, which is housed in the District Magistrate's Office in the district headquarter, was under the control of the Trinamul Congress. The Parishad had 73 elected seats out of which the TMC had got 34 (and formed the Board), the CPI-M 26, the RSP and Socialist Unity Centre of India (SUCI) each having 5 seats and the Congress 3 (as per 2009 panchayat election results).

[13] Under the Trinamul-led government the Minister-in-charge of the irrigation department in his budget speeches focused on the reconstruction of the Aila embankments (Bhunia 2011, 2012). The minister presented land acquisition proposals for the purpose of rebuilding embankments. The minister stated that land acquisition proposals involving a total of 5,935.48 acres were submitted to the district land acquisition collectors between February and July 2010 (Bhunia 2011). Similar land acquisition proposals were also made in the budget speeches of 2012.

4

Treading a Fine Path between River and Land
Livelihoods Around Embankment

Island, stakeholders and location

As development thinking has moved from 'thing-oriented' to 'people-oriented' paradigm (Chambers 1995),[1] it has thrown up a number of terms such as 'poor' (World Bank 1996a), 'poorest of the poor' (Harris 1988; Singh 1988), 'indigenous people' or 'communities' (United Nations 1993), 'beneficiaries' (Paul 1987; World Bank 1996b), 'stakeholders' (Nelson and Wright 1995; World Bank 1996b; Jeffery and Sundar 1999; Vedeld 2001) and so on to refer to people. The 'poor' are defined as people living in remote and impoverished areas (World Bank 1996a, 7). The term 'stakeholders' has been further classified into 'primary stakeholders' referring to people or direct beneficiaries and 'secondary stakeholders' including government agencies and NGOs (Vedeld 2001, 6). A question that arises is who these 'stakeholders' or 'poorest of the poor' are. How does one understand these meta-categories in a particular context? The chapter is an attempt to answer these questions.

Like other southernmost Sundarbans islands, Kusumpur also has mud quays or embankments encircling and protecting the islands. A part of Kusumpur island is connected with the landmass of Gosaba island that comes under Gosaba block and constitutes a separate village panchayat. A road from Gosaba market leads to Kusumpur and then branches off into two directions, one going in the direction of Jagatpur and the other leading through south Kusumpur towards a van rickshaw stand (where van rickshaws are parked) next to Kusumpur High School. Beyond this van rickshaw stand are the winding embankments serving as pathways connecting Jaipur, Ramnagar and heavily eroded north Kusumpur to south Kusumpur. These embankment paths in most parts of the island are so narrow that van rickshaws cannot ply on them. People mostly walk or use bicycle to travel to their destinations. There are stretches where embankments become so narrow and breached that even cycling proves difficult. People get off their cycles and walk through that particular stretch of embankment path carrying the cycles on their shoulders. Embankments are always higher than the ground level of the inhabited islands to protect them against flooding during high tide when water rises higher than the islands. Embankments, when viewed from inside the villages, look like mud walls with people using the slope of the wall to ascend and descend.

Branching off in horizontal lines from the embankment at regular intervals elevated mud tracks pass through the interior of the island. Sometimes mud tracks are intersected by brick-paved paths connecting villages and neighbourhoods within the island. Jaipur and Ramnagar villages are internally connected by mud paths stretching in a more or less horizontal line along the embankment. Because of continuous erosion and shrinking of space in north Kusumpur, Ramnagar and north Kusumpur are connected only by embankments surrounding both the villages. However, mud tracks or brick-paved paths are not unique to Kusumpur alone. In most of the Sundarbans islands, mud or brick-paved paths inside the islands are meant to cover short distances, but to traverse longer distance embankments remain the only pathways. Houses of those who are economically well off and own lands are generally located towards the safe interior of the islands, whereas poor families – particularly those working largely as labourers or engaged in prawn seed collection or prawn trade – have their houses located on the periphery of the island, scattered along the embankment and exposed to the risks of embankment collapse and being swept away by tidal waves. Having a house in the middle of the island does not necessarily ensure safety because saline water breaking through the embankment fast engulfs houses, ponds and fields, but still people living in the interior of the island are less vulnerable than those living on the periphery.

One's location in the island also provides a clue to the sequence of migration. As has been mentioned earlier, Kusumpur island had been populated by migrants from Bangladesh and Medinipur in West Bengal. Kusumpur village was populated entirely by the people from Medinipur. Only the western part of the village was inhabited by tribal families. However, people from Medinipur were not simply confined to Kusumpur village alone. They settled in villages of Ramnagar and Jagatpur as well. Because they settled first, most of their houses were found located in the interior of these two villages. Houses of the Bangladeshi migrants in these two villages were located on the path along the embankment. Jaipur village was settled largely by the Bangladeshi migrants. The families which migrated early in the 1940s and 1950s managed to have their land and houses away from the embankment towards the interior of the village. But families which migrated much later did not have a choice and settled closer to the embankment.

However, when one turns to the eroded stretch of north Kusumpur, the place where about 130 families lived under the threat of a displacement (see Map 3), one realizes that the Sundarbans' ecology had been a great leveller. The families who lived here came from Medinipur in the early 1940s. They had their land and houses located towards the safe interior of the island. However, tidal waves, continuous land acquisition and new embankments had done away with their notions of interior and periphery. Out of these 130 families, only 15–20 families had been economically well off and bought land in the vicinity of Kolkata. The male members of these families were either primary school teachers or clerks in government offices. The remaining 110 families were fast losing the land beneath their feet. Thus, here we come across a community of islanders (or stakeholders) who shared a common destiny. However, this community became visible in its endless negotiations with the governmental machinery, particularly the machinery of the irrigation department and its activities around embankments in north Kusumpur. Instead of viewing communities along the lines of caste, tribe or religion, the chapter views communities as being produced by two interlinked processes: (1) by the eroding land and landscape of the delta; and (2) by specific instances of governmental interventions aiming at embankment protection and disaster control. People living in this narrow and eroded patch of land are fighting displacement with their back to the wall. Their sense of vulnerability clubs them into a single category (a community of islanders constantly backtracking as they lose ground). Here, I will digress a little and situate the specific empirical case of north Kusumpur in a wider historical context, delving into the colonial riparian rules and the role it played in producing vulnerable landscape and settlement elsewhere in the Gangetic delta. I will also show that in a vulnerable landscape community ties tend to be contingent in nature. A collapsing or collapsed embankment entails displacement and misery for the islanders, but paradoxically, building that embankment presents them with the possibility of a livelihood. It is in search of livelihoods that islanders engage the governmental machinery meant for embankment construction and flood control. The chapter documents some of the negotiations that the islanders have with the irrigation department. The chapter highlights different strategies that these islanders (also stakeholders) in the capacity of embankment labourers or labour supervisors (dafadars) resort to, ventilating their grievances or subverting the powerful governmental machinery. The chapter also turns to the strategy that the displaced islanders of Garantala island adopted for voicing their protest against governmental inaction. Finally, the chapter revisits the 'modes of protest' or 'weapons of the weak' theories in the light of the empirical evidence from the Sundarbans.

Eurocentric riparian doctrine and settlement issues in the Gangetic delta

In a state of actual or perceived vulnerability, villagers come together as was the case with settlers of north Kusumpur. A question that arises at this point is whether, in viewing communities as being produced by the peculiar ecology of the region, one runs the risk of being accused of environmental determinism. To provide a categorical answer to this question proves difficult. We need to look at the question in a historical perspective, a perspective that delves into the problems entailed in trying to settle unstable lands. Chapter 2 looks into the reclamation history when colonial rule had the Sundarbans wetlands prepared for settlement and revenue generation. The chapter has looked at the problems associated with reclamation and subsequent change in the colonial settlement policy in the Sundarbans. However, our discussion here is not restricted only to the Sundarbans. I would like to widen our focus and make the Sundarbans part of the larger problems associated with the Bengal Presidency.

In talking about laws relating to water rights and ownership of lands bordering rivers, Hill[2] notes that the social and ecological history of colonial India was deeply affected by the importation of riparian doctrine (1990, 1–2). Colonial rule tended to universalize the riparian doctrine modelled entirely on European environmental concerns. The riparian legislation evolved primarily due to environmental factors with an average rainfall being a crucial agent in the development of water rights. In the Bengal Presidency where some areas can experience an average rainfall of over one hundred inches, the Eurocentric riparian laws were found inadequate in dealing with the destructive properties of the Bengal river systems (Ibid, 2). The Ganges, Brahmaputra and the Kosi were some of the rivers in the Bengal Presidency that had the power to pose a threat to and devastate settlements in their surrounding areas. Most of the rivers originate from the Himalayas and carry enormous sand or silt. The continuous deposition of silt coupled with heavy rain during monsoon caused these rivers to flood the adjacent regions rendering them unsuitable for human habitation and settlement. Reflecting on the nature and course of these rivers, Hill argues, 'The land that the rivers leave in their wake, as well as the land bordering the new beds that they form, is officially known as a "fluctuating river tract," or *diara*' (Ibid, 3; italics as in original). Illustrating the case of the Kosi,[3] one of the most violent rivers in the Gangetic delta, Hill states,

> The Kosi has historically traced a direct and violent path to the Ganges, passing through the Chatra Gorge in Nepal, where the valley is only three to five miles in width; this increases the velocity and power of the river ..., the sand within this furiously flowing river is not ejected until it nears the Kosi-Gangetic conjunct. Accordingly, the diara surrounding the Kosi is primarily composed of sand, leaving the land uncultivable for years ... Diara is further of a very temporary nature, shifting from year to year. In many, indeed if not most situations, land that is eminently cultivable one cropping season is under water the next; ... For centuries, these riverine tracts have been the focus of prosperity one year and tragedy the next, overpopulated one decade and deserted for the following generation. The violent characteristics of the Kosi river have had powerful social ramifications, involving desertions, migration and agrarian exploitation. For the newly-empowered British East India Company, however, no problem was more baffling than revenue collection in the diara estates (Ibid, 3–4).

In contrasting the pre-colonial Kosi diara region with the colonial, Hill argues that before the coming of the British, the region did not pose any great difficulties to revenue collection and productive capabilities. This was largely because the region was sparsely populated and the indigenous administration was flexible in adjusting its revenue demand to suit the vagaries of environment (Ibid, 4). However, it was the colonial revenue and settlement policy that drastically altered the pre-existing social and political equation in the region. Colonial rule instituted land and water rights modelled on the European riparian legal framework. Thus, it is the imperialism of categories (Nandy 1990, 69) that allows specific European experience and the knowledge born of such experience to be applied universally. It was as though European science and legal principles could be applied anywhere in the world in complete disregard of specific histories and ecologies of a region.

Hill argues that the riparian doctrine might have been suitable for pastoral and industrial England where rivers were relatively constant. Therefore, the law provided a legal recourse to riparian owners damaged by the growing industries in England. The model, developed largely in response to similar environmental realities of England, western Europe or eastern United States of America, could have been a better safeguard against loss of land to flooding or possibility of losing sources of water (Hill 1990, 7). However, applying this doctrine to the Indian context, particularly to north-eastern India where rivers were turbulent and changing their courses dangerously, had been disastrous for people and their lives. The British colonial policy of settlement and revenue generation went unabated. And what contributed to the consolidation of this process was the Permanent Settlement Act in Bengal which created a 'class of an Indian equivalent of the English yeoman-farmer' (Ibid, 7) who would demonstrate their revenue loyalty to the empire.

Colonial authority completely ignored the problems involved in making extremely unstable land a source of stable revenue. We know how the Permanent Settlement worked with the zamindars being at the helm of affairs and the rayats or peasants at the bottom of the hierarchy. The peasants were oppressed and tortured for revenue regardless of whether lands had produced enough. According to Hill, the problem was acute in the Kosi diara region.

> Peasants were deserting their holdings rather than face prosecution for inability to pay the assessment on inundated lands. If they stayed during times of abundance, they found their plots being sold from under them when the land, which had been so productive the year before, was no longer fit to fulfill the terms of their contracts ... No provisions had been introduced concerning proprietary rights to the old beds of the river. Who owned this emerging land, and what rights to ownership did the landlords have over that land now inundated by the Kosi (Ibid, 9)?

Several acts were passed and surveyors deployed to address the problems of diara land and to follow a more stringent revenue policy. As a consequence, the peasants remained deprived and alienated. Even the Bengal Tenancy Act did not improve the situation for the better because the act decided to grant occupancy rights to a peasant if he had held a diara land for twelve continuous years (Ibid, 15). Colonial rule did not even consider it necessary to raise the question of whether a peasant could hold on to a land, so transient and unstable, for twelve years. The colonial stringent revenue policy coupled with oppression, extraction and eviction of peasants by the landlords resulted in violent protest movements by 'the diara sharecroppers, mostly tribal Santals from South Bihar' (Ibid, 17).

Colonial rule had always been uncomfortable with those who did not practice settled agriculture. Just as colonial rule faced problems in implementing its settlement policies on unstable diara tracts, similarly the rule encountered problems in making the Santhals submit to settled agriculture. The Santhals' way of life, which focused largely on 'communal organization and wet rice cultivation, hunting and forestry' (Hill 1991, 273), posed a potential threat to the British empire, which looked upon settled agriculture as the source of revenue generation. There were efforts on the part of the British authority to bring the Santhals within the fold of settled agriculture. After the Permanent Settlement, the Santhals found their forest-based life challenged by the newly imposed British law (Duyker 1987, 159). Their entry into the world of agriculture resulted in their being in perpetual debt to money lenders followed by confiscation and illegal eviction from the land (Hill 1991, 273). Nineteenth and twentieth centuries witnessed many Santhal rebellions, which were expressions of tribals revolting against the landlords and the colonial state. The image of Santhals as robust and rustic manual labour deployed by colonial labour market gained considerable ground (I will focus more on this in the next chapter where I discuss embankment, forest and stereotyping of adivasis). Despite the formation of a separate district of Santhal Parganas, Santhals were not kept confined to their district. The Santhals, constrained by the demands of colonial labour market, had to migrate to different parts of India to reclaim lands lost either to water or to forest. The Santhals' past as hunters and forest dwellers came in handy for the commercial expansion of colonial rule. However, the Santhals' exposure to settled agriculture led to their impoverishment and deprivation. Duyker (1987) shows how this sense of deprivation among the Santhals led them to join the Naxalite movement in Bengal in the 1970s and launch a guerrilla war against the Indian state. The Santhal rebellion figured prominently in other scholarly writings (Singh 1966, 1983) and also in Bengali fictions such as Mahasweta Devi's *Aranyer Adhikar* (Devi 1977) (Right to Forests).

The Sundarbans wetlands, transient, fluid and connected by rivers and rivulets, where land surfaces only to disappear, posed a serious obstacle to the land hungry policies of colonial rule. Like most of the rivers of the Gangetic delta, the rivers in the Sundarbans are frequently changing their courses. Furthermore, the region's proximity to the sea is what makes it ecologically fragile and vulnerable. Yet, colonial rule, as mentioned earlier, was intent on turning the unsettled delta into settled land for revenue. Unlike the Kosi diara region, the Sundarbans delta could not be brought under the Permanent Settlement Act, for land surveying and boundary making proved even more difficult. Still interventions continued in the region shaping and determining ecologies from the commercial viewpoint of colonial rule. However, a significant question that arises at this point is what riparian rules could have guided colonial settlement policies in a region where land bordering water was non-enduring and rivers were so saline that water rights even for the purposes of irrigation proved wholly ineffective. Land rights were simply vested in people, but very scant attention was paid to the implications such rights would have for lives people lived in the delta.

Even today no legal recourse is available to people in the event of their land being lost to rivers or to new embankments built, which acquire more lands. It is curious to note how lands are lost not only to rivers but to lands as well. I have already mentioned how acquisition of land from inside the village is a precondition for building a new stretch of embankment. The villagers lose their lands and in the absence of any legal recourse, stand reconciled to the loss. The embankment protects cultivable lands against salinity in the rivers. But when a new stretch of embankment is built inside the village behind the old, broken or damaged stretch, the land that is acquired for the new construction now falls outside the village and becomes part of the river bank. People's right to this land is inconsequential, for salinity of the river makes it uncultivable. The only solution lies in turning this land into a fishery where brackish water prawn farming is possible (more on prawn farming and politics in Chapter 6). Prawn catching and farming is an established source of livelihood for people in the region. But it is also true that for setting up a fishery a reasonably large land area is required. The villagers of north Kusumpur, on whom this chapter focuses, are victims of land erosion. The village is shrinking so rapidly that people do not have adequate land available to set up a fishery. Thus, in looking at these villagers on the threshold of land loss and displacement, one tends to be environmentally deterministic. Our discussion above amply explains why one becomes deterministic. One encounters conditions of marginality induced not simply by fragile ecology, but by the absence of legal recourse to address one's vulnerability. The subsequent sections of this chapter document what people in a state of vulnerability do to earn their living either individually or as a collective.

Building whose embankments: Instances of marginality and agency

Most of the construction and repair work is undertaken by the irrigation department during the monsoon months when embankments are particularly vulnerable and prone to collapse. Needless to say that heavily eroded north Kusumpur was a site for irrigation's embankment-related work. During my stay in Kusumpur, I was a regular visitor to north Kusumpur where construction and repair work used to be undertaken by the department. Ironically, embankments which spelt disaster for the people could also be seen as a source of livelihood for the islanders. And more often than not the department employed the villagers as labourers for construction and protection of the embankment. Thus, the villagers of north Kusumpur who were fighting land erosion found in this construction work a source of earning for their families as well. Carefully maintaining my balance on the slippery and muddy embankment, whenever I reached the construction site in the morning, I would find people already engaged in their day's work. There were some working on the slopes of the damaged stretch of the old embankment, some digging earth from the nearby fields or riverbanks, some carrying earth to add to the new embankment under construction behind the damaged one and yet others trying to steal time away from their work to smoke.

On one such day when I reached the construction site along the bank of the river Goira in north Kusumpur I found the workers waiting for some bamboo poles to arrive on the irrigation department boat. When finally they did arrive, a number of disputes arose with respect to their quality. One of the labourers ran down the slippery slopes of the embankment to where the boat was anchored and then shouted to his colleagues at work, 'These bamboos are no good, they will not last long'. Those standing on the embankment judging the quality of the bamboos from a distance shouted back, 'It seems that they have not sent the number we have asked for. How many bamboos are there?' They asked the irrigation employee (khalashi)[4] standing on the deck of the boat. 'There are about seventy bamboos here', he answered. 'But we asked for at least one hundred bamboos, what are we going to do with seventy?' They discussed the matter with each other and then, turning to the employee said, 'Take the bamboos back and tell the irrigation people that we are not going to work with seventy bamboos'. The employee promptly answered, 'This is what we have managed to get. If you cannot work with these it's your problem'. He then turned to Dinu, the dafadar sitting next to the workers and asked him, 'Why don't you explain it to them?' Dinu tried to reason with the workers saying, 'Start your work with seventy bamboos and we can ask for more later. But if you send this consignment back then the work will be stopped. The irrigation SO [section officer], SDO [sub-divisional officer] and panchayat will then intervene and the situation will go beyond our control'. He spoke quietly to ensure that only the workers could hear him and not the khalashi standing at a distance. Dinu patted the shoulder of one of the workers who was the most vocal of all and said in a patronizing way, 'Why ask for trouble?' It took Dinu some time before the workers finally agreed to work with only seventy bamboos.

Soon after the bamboos were unloaded and the boat had disappeared round the bend of the river, Dinu heaved a sigh of relief and took a second look at the bamboos. Smiling at me he said, 'The bottoms are okay but towards the top the bamboos are weak'. I asked Dinu about the use to which these bamboos would be put. Pointing to a pond behind a stretch of the newly built embankment, Dinu said that the bamboos would be required for reclaiming a portion of the pond. This reclaimed part would then provide support to the new embankment. It would protect the embankment against water pressure during high tide and would prevent the pond from destroying the bottom of the new embankment. If the pond was not adequately reclaimed, there would be a possibility of earth of the embankment melting away into the pond.

At the insistence of Dinu, the labourers started cutting each of the twenty feet long bamboos into equal halves of about ten feet each. While this was happening two workers went into the pond and started positioning a couple of bamboos across the breadth of the pond at regular intervals. One held each piece of bamboo in an upright position and the other hammered it into the bed of the pond to ensure that half the bamboo went under the water and the rest remained above the water surface. After the requisite number of bamboos had been placed in a row, a long piece of bamboo was vertically sliced into halves, which were then nailed laterally onto both sides of the bamboo pieces half immersed into the water, in such a way as to keep the bamboos in a single alignment. The remaining bamboo poles were then carefully positioned next to each other inside the channel and hammered down firmly into the bed of the pond. Thus, the structure now looked like a wall made up of bamboos half immersed in water. Finally, a huge sheet made of bamboo fibres was placed against this wall to plug whatever gaps existed between the bamboos so as to prevent water seepage from the non-reclaimed part of the pond into the reclaimed part. Once the structure was complete, reclamation started. The labourers poured earth into the portion of the pond demarcated by the structure.

Dinu as dafadar

During the entire process of reclamation, Dinu, whom I introduced in Chapter 3, was supervising the labourers while chatting with me. It is interesting to know how Dinu managed the dafadar's job. He shared with me this fact when I met him in his house. Like many others he lived on the narrow stretch of land between the Matla and Goira rivers. Dinu's house was in bad shape and severe cracks had appeared on the mud walls. Pointing to the cracks I asked him, 'Are these due to the salinity in the soil?' 'Partly due to salinity and partly due to lack of maintenance, no money', he answered.

'Why, now as a dafadar you must be earning a lot?' I asked. 'Well, I needed this job', he replied and continued, 'But you know how I got it? First in order to get a certificate from the party-panchayat,[5] I had to give them a bottle of *foreign liquor* [whisky] and two chickens for dinner. Then I had to go to the irrigation office at Gosaba for their approval. There they asked for two bottles, as one was not enough. And on top of all this, I had to run errands for the engineers for a month. I used to go to Gosaba irrigation office at ten in the morning, sit there and be at their beck and call the whole day and come back to Kusumpur at six in the evening'. 'Why did you take so much trouble, when you could have worked as a labourer and earned even more?' I asked in surprise. 'Oh no, you have not understood how labourers are paid', Dinu replied. He tried to explain, 'The money that labourers get is shared amongst ten, fifteen or may be twenty of them whereas whatever I get is mine. I need this money. I lost all my land due to earlier ring embankment constructions. These engineers, contractors, party-panchayats are all making money. What's the harm if I make some too?'

Dinu's narrative makes it clear why he tried to pacify the labourers when they complained about the quality of the bamboos. Dinu desperately wanted the supervisor's job and he would seek to retain it at any cost. The way he handled the dispute shows that he was cautious and wanted to play safe in his dealings with the irrigation people. Under no circumstances would he antagonize the irrigation engineers and contractors, lest he run the risk of losing his job. However, this does not mean that Dinu was not sympathetic to the cause of his fellow workers who were also his neighbours. With them he had shared many anxious moments and survived four ring embankments. Like many others he too lost a substantial tract of agricultural land. The pond they were reclaiming belonged to Judhistir, who happened to be Dinu's neighbour. Like Dinu, Judhistir had also lost land due to earlier ring embankments. His agricultural land of less than a bigha included this pond as well. He constituted a part of the labour force and ironically he himself carried earth from the riverbank the whole day to reclaim his own pond.

Embankment workers

Every time land erodes or an embankment collapses, it amounts to loss of private space, a space the villagers call their own. However, this sense of loss does not condemn them to a state of utter despair. Rather, it imbues them with a sense of urgency that has found expression in a variety of subversive acts. If contractors cheat labourers by not giving them their due payments or make profits by supplying bamboos or bricks of inferior quality, the latter in turn make up for their loss by selling the contractor's property such as bamboos, bricks and so on. They even deceive the contractors by not digging the amount of earth proportionate to their wages. In popular parlance this is often referred to as 'stealing of earth'. I always found it difficult to understand how earth was stolen. Although I spent considerable time in the company of the labourers, I felt hesitant to pose this question. Finally, I decided to single out Kalyan and Shankar, both residents of north Kusumpur, with whom I had developed some rapport. On posing the question, I noticed frowns on their foreheads, which soon disappeared when both started laughing and asked me in return, 'Tell us something, is that also relevant to your research?' 'Yes, to a certain extent, but then I am also curious to know how earth is stolen. But if you don't want to tell me, I won't press you', I assured them further. To my surprise both drew near me and said, 'It is not a question of us divulging a deep secret. We all do it. But do not tell others that we showed you'.

They took me to the riverside. Pointing to the square holes (chouko) on the riverbank one of them said, 'We make these holes to dig up earth for the bund [embankment]. Each of these square holes has to be ten feet in length, breadth and depth. We are paid per thousand cubic feet of earth dug up. Now most of these holes are inundated during high tide and remain filled with water even when the water recedes during low tide. When construction work is undertaken afresh by the contractors we simply increase the length and breadth of the existing holes but we do not do them that deep. Because they are filled with water one cannot measure the actual depth of these holes'. Shankar looked at me and smiled. 'If they can cheat us why can't we cheat them?' 'But don't you think by deceiving the contractors and engineers you are weakening the embankments and endangering your own existence?' I asked Shankar. He replied, 'In any case there is not much to lose. We don't have much ground left beneath our feet, do we, having survived four ring bunds? It hardly matters if there is one more. Everybody makes money, what's the harm if we make some too?'

The displaced in Garantala and their protest

At this point, let us turn to Garantala island where the villagers lost their lands because of the 1,500-feet wide ring embankment (see Chapter 3). As mentioned earlier, there were about ninety families who lost their lands and of these ninety families about thirty lost their houses as well. These thirty families lived closer to the damaged embankment and therefore could not save their houses. After being rescued by their neighbours, they managed to build temporary makeshift houses on the embankment, a little away from the stretch where the embankment caved in.

These people were given to understand by the village and block panchayats that soon after the protective ring embankment was constructed they would be relocated elsewhere on the island. However, even after the construction of the 1,500-feet wide embankment, which acquired more than 36 acres of land, these thirty families were made to continue where they had set up their temporary houses. Despite their repeated visits to the local and block panchayat offices their situation did not improve. Even the victims' appeals for compensation was ignored by the panchayat members on grounds that since these ninety families already turned the 36 acres of their lost land into a fishery[6] and started to earn from it, they would no longer be in need of compensation. Almost six/seven months passed and no help seemed to come their way. Meanwhile the villagers decided to pursue their cause. By paying a surveyor at the local land records office, they managed to procure a map of the 36 acres of their lost lands and visited the Land Assessment Collectorate in the district headquarter in search of compensation. They waited in vain in the corridors of power, got exposed to the complex legal procedures and also met middlemen (dalal) who proposed to act on their behalf in return for payment, but justice still remained elusive.

'Now it was time for us to pull up our socks', said Ananta Mistry during my interview with him and others who were now settled in their new land in Charkhali on Garantala island. Charkhali was a government land lying to the south of Garantala island. It was largely a vacant land which had only a few families, migrants from Bangladesh, who were settled by the local RSP (which had a strong electoral base in Gosaba). On being denied justice and realizing that no help would come their way, Ananta and others, who lived on the embankment for nearly seven months, decided to defy their fate. With the help of their friends and neighbours, these thirty families set out at night and traversed a long winding path through the rice fields before they finally reached Charkhali and forcibly occupied the government land. They fenced their occupied land with bamboo poles as a mark of their protest against governmental injustice. However, the local RSP panchayat set its face against such forcible occupation of land and next morning the RSP panchayat members demanded that the settlers evacuate the occupied land. According to Ananta and Ramen Baidya (another settler in Charkhali), the RSP leaders were almost threatening in their demeanour. But the settlers held on to their ground. They were supported by the local CPI-M leaders who, I came to know, stood by these families in times of need and supported their initiative to settle in Charkhali. It is perhaps needless to say that by providing support to these families the CPI-M wanted to discredit the RSP and expand its base in an area where the RSP reigned supreme. But then, the CPI-M's political interests converged with the dire needs of these families, and in this case, this convergence of interests worked to the advantage of the settlers.

Post-Aila scenario

I went to the Sundarbans three months after the cyclone Aila. On my way to Kusumpur, I travelled through some of the worst affected villages in Basanti and Kultali blocks (blocks of the South 24 Parganas part of the Sundarbans). The villagers, a majority of them, had their houses razed to the ground. The cyclone had robbed people of their sources of livelihood. They did not have fish in their ponds, rice in the fields and those surviving as wage labourers suddenly found that the sources of employment had dried up. It was as if the economy had come to a halt in the villages in the Sundarbans (Mukhopadhyay 2011, 21). As a result, people left the Sundarbans in search of livelihood elsewhere in India (Ibid). Agriculture remained non-functional and ceased to have become a source of livelihood in the villages. The villages were waterlogged for more than a month after the cyclone. In many villages, the rice fields were still found saline-water logged.

Kusumpur also wore a deserted look. Rice fields in Kusumpur looked pale and devastated. In north Kusumpur, majority of the houses were completely or partially destroyed. Out of 130, only 10 houses survived the cyclone's onslaught. The families lived in makeshift houses, which had mud walls (two to three feet high from the ground) and polythene sheets to cover the roof. During Aila saline ingress contaminated the ponds which were the only sources of freshwater. Most of the families in north Kusumpur could not desalinate their ponds because they did not have enough money. I met the residents of north Kusumpur who stated that they did not have rice for nearly two months after the cyclone. For the majority of the families, the staple diet was boiled flour made into a thick paste which they sometimes had with boiled potatoes. For many families, sources of livelihood had simply dried up. Before Aila, landless wage labourers' earning was in the range of Rs. 100–150 per day, but Aila had reduced this to a mere Rs. 50–60 per day (Ibid).

I visited Kusumpur again in 2010, a year after the Aila, but there had been no substantial change in people's quality of life. Rice fields wore a barren look. Rice cultivation failed immediately after the Aila, even after a year, the situation had not improved. Considering salinity in the soil, people in Kusumpur tried high-yielding variety seeds (Pankaj) in 2010, but lands which were water logged for more than a month failed to yield any rice (Ibid, 22). And even in lands where rice grew, the yield was abysmal. The question of a second winter crop did not arise because of the absence of freshwater. People with a bigha of land could produce about 60–70 kilograms of rice, whereas the pre-Aila yield was about 600 kilograms. Villagers I met stated that the present yield was unsustainable considering the investment they made. For a bigha of land, farmers needed to buy 15 kilograms of seed which cost about Rs. 150. To plough the land, they had to hire a power tiller which cost Rs. 380 per hour. In addition to these, farmers needed to employ labourers at the time of sowing and harvesting at the rate of Rs. 120 per labourer per day (Ibid).

The plight of the families living in the eroding stretch of north Kusumpur was deplorable. Many of the families had not been able to rebuild their damaged houses. The houses still had polythene sheets serving as roof tops. Immediately after the cyclone, the government declared Rs. 2,500 and Rs. 10,000 as compensation for partially and fully damaged mud houses, respectively (Ghosh 2010, 119). Even after a year, none of the families in Kusumpur had received any compensation from the government. North Kusumpur looked quite deserted. I could not find the people I knew because the majority of the houses (almost 80 out of 130 families) did not have their male members. People like Kalyan, Shankar, Tapan, Chandan and Bhagirath, whom I already introduced here and in the earlier chapters, had left their families in search of work in Chennai and Andaman and Nicobar islands. In most of the villages of Gosaba, Patharpratima or Kultali blocks in South 24 Parganas or of Sandeshkhali blocks in North 24 Parganas, young and middle-aged people had left their households. This had a negative impact on the economy in the region as agriculture ceased to become a meaningful activity. Ever since the cyclone, prawn farming has been on the increase and it has become a major source of income for many families. As a consequence, the number of tiger prawn fisheries (or locally called bheri) has increased steadily.

With the majority of young villagers having migrated, north Kusumpur looked much different from what it used to be when I spent time in the company of the embankment workers. There were not many known faces around. The embankment stretching through north Kusumpur looked as though it was in need of urgent repair, but villagers were no longer found at the embankment site. However, I met Dinu who was found sitting in front of his house and talking to Barun Patra, one of his neighbours. To my surprise I found that Dinu had his house rebuilt and the roof thatched with an asbestos sheet. During my visit to Kusumpur immediately after the Aila I had found Dinu's house badly damaged. I remembered very clearly that his house did not have a roof and a portion of the mud wall had completely collapsed. Seeing me both Dinu and Barun smiled and Dinu asked, 'When did you come?' 'Yesterday', I answered. On exchanging pleasantries, Barun left as he had to go to Gosaba market. I followed Dinu into his house. I sat on a plastic chair while Dinu sat on a cot lined against the wall. 'So how is it that you managed to rebuild your house, while your neighbours have failed to build theirs?' 'Is it only you who have got compensation?' I asked. 'No, none of us got compensation, we are all awaiting it. I rebuilt the house with the money I got from my supervisory work [the work of a dafadar]. Ever since the cyclone, embankments are in bad shape, they are prone to collapse in many places. The irrigation has been engaged in repair and rebuilding of embankments. I had been employed as supervisor in some of the places in Gosaba, Kusumpur and Bali [islands of Gosaba block] where repair work took place', Dinu answered. 'So fortune favours the brave, good, you did not go to Chennai or Andaman in search of employment', I smiled. 'I could never have gone to Andaman leaving my family behind. My son is only seven years old, my wife alone won't be able to manage', Dinu stated emphatically. 'Yes, I understand your predicament. I hope you will have more supervisory work coming your way', I tried to pacify him. 'But I don't know what would happen to my job, because now I have a competitor', Dinu looked a bit pensive. 'All this while I was alone, but now Srinath has also become a dafadar. You know Srinath, don't you?' Dinu asked impatiently. I knew Srinath who lived in north Kusumpur, a few houses away from Dinu. 'Yes, I know Srinath. He used to work as a construction labourer', I responded. 'Now he has become a supervisor and these days he is spending a lot of time at the irrigation office in Gosaba. He is always found at the irrigation office running errands for the engineers', Dinu stated. 'It is true that Srinath is doing all that to earn a living, but does that mean that your earning will stop?' I asked. 'You should go to Gosaba irrigation office and see what Srinath does throughout the day.

Every day he brings for the engineers vegetables he grows in front of his house. He buys them chicken and *foreign liquor*, he invites the Gosaba and Kusumpur village panchayat heads to dinner whenever he cooks chicken for the engineers at the irrigation office. Recently, I went to the irrigation office to enquire if the engineers would consider me for a supervisor's job for an embankment repair work coming up in Kusumpur. You should have seen how dismissive the engineers were, how badly they treated me', Dinu looked crestfallen.

In a muddy and slippery terrain where everything looks so transient, relations tend to become contingent and fleeting. Dafadar's job became a bone of contention between Dinu and Srinath. In a place where opportunities of livelihood are difficult to come by and earning a livelihood is a daunting task, people do not let go anything that comes up their alley. Dinu's narrative mentioned earlier in the chapter makes it clear how he managed the dafadar's job. Dinu was saddened at the prospect of losing his job to Srinath who had exactly followed Dinu's footsteps in becoming a dafadar. It was as if by criticizing Srinath, Dinu reminded himself of how he once became a dafadar. Dinu's sadness deepened because Srinath happened to be Dinu's neighbour. As neighbours they lived on the same erosion-prone stretch of north Kusumpur and shared a common destiny. Yet, as dafadars they tended to outwit each other in grabbing opportunities.

As mentioned at the beginning of the chapter, communities are also produced by specific instances of governmental interventions aiming at embankment protection and disaster control. By governmental interventions, I meant irrigation department's embankment-related activities. It is in the context of these activities that we witness villagers' role and experiences as embankment workers and labour supervisors. There is no denying the fact that corruption figures as a recurrent theme in the activities and experiences documented above. Corruption, presented here more as a blame game, should not be viewed as dysfunctional. Rather it is a lens that allows us to see specific instances of development and their mundane and everyday character. Narratives of corruption help us see how the villagers participate as workers in the governmental machinery of development and seek to achieve, even in a state of crisis and vulnerability, their individual or collective aspirations. I have discussed how the labourers collectively steal earth or irrigation department's property. I have also discussed Dinu or Srinath's endeavour to become a dafadar. These narratives are significant in that they lend visibility to governmental development.

Revisiting everyday forms of protest

The above discussion presents us with interesting facts about people's resistance. Despite being victims of governmental injustice, the strategies the islanders resort to for ventilating their dissent are many and wide ranging. There are dafadars like Dinu and Srinath who wanted to survive within the system and make the most of it. They are part of the labour force and also a representative of the irrigation office in Gosaba. Dinu feigns ignorance when labourers sell the irrigation department's property, but is equally alert to the task of not letting such activities assume proportions where his position as a dafadar is threatened. The strategies of Kalyan, Shankar and their colleagues who work on the construction site are clearly subversive. The strategies such as selling of contractor's property or stealing of earth initiate a process whereby the limits of the order established by the powerful is tested. They prefer to outwit the engineers and contractors at their own game. These strategies are used to ventilate grievances and find a breathing space in a place like the Sundarbans where living exposes them to enormous risk and vulnerability. And at the end of the spectrum are the villagers of Garantala who forcibly occupied government land as a mark of their protest against governmental apathy and neglect.

Looking at these different modes of protest or dissent, one is tempted to use Scott's much celebrated thesis on the forms of resistance (Scott 1985). On the basis of his study, Scott argues that showing disrespect or using slander, gossip, theft and so on are various ways in which the poor or subordinates resist the dominance of the powerful. According to Scott, the dissent registered in most cases is mainly symbolic, but such symbolic dissent continually presses against the limit of what is permitted on stage, much as a body of water might press against a dam. In this pressure, for Scott, lies the desire to give unbridled expression to the sentiments voiced in the hidden transcript directly to the dominant (Scott 1990, 196). Thus, the weapons, the so-called weak use, initiate a process whereby the limits of the order established by the powerful is constantly tested. Scott observes this process in a peasant society where dominance of the rich is often subverted by recourse to such strategies as mentioned above. Later Scott's perspective was used by Haynes and Prakash (1991) to understand the 'everydayness' (Ibid, 16) of resistance not only in production relations but also with regard to popular culture and gender in South Asia. Their volume contains reflections on the lifestyle of the courtesans of Lucknow in India (Oldenburg 1991) and on gambling in public spaces in colonial Sri Lanka (Rogers 1991), which clearly reveal everyday strategies of resistance.

Scott's perspective is significant in that it helps us understand Kalyan, Srinath, Shankar and their ways of living and surviving in the Sundarbans. Here too we are confronted with theft and non-compliance reflected in the stealing of bamboos or bricks. These are the rituals practised to ventilate grievances, earn a living and find a breathing space in a life that otherwise looks bleak. However, Scott's approach has not gone unchallenged. His theory of passive or symbolic resistance is criticized on grounds that such acts of subversion do not alter the balance of power in favour of the oppressed. In other words, symbolic resistances as described above do not turn the existing order established by the powerful upside down. Some have also argued that Scott's formulation does not help us understand why certain forms of petty thieving are looked down upon universally regardless of class differences, whereas people have no problem in romanticizing a Robin Hood who steals from the rich or powerful to give to the poor (Gupta 1997, 116).

Apart from these criticisms, Scott's theory is also problematic at another level. Linked to his notion of subversion or resistance is the implicit notion of solidarity. If the powerful dominate their subordinates as a group, the latter are also made to appear as if they subvert as an organized force, even when their resistance is sporadic and eclectic. Scott suggests that an act of insubordination, when not rebuked or punished, is likely to encourage others to venture further and the process can then escalate rapidly (Scott 1990, 196). Therefore, the success of a subversive act lies, to a considerable extent, in its being a shared reality and to the extent it is shared, it can organize the practitioners into a collectivity. It is in this sense, the weak or poor or subordinate, for Scott, form a homogeneous group. The successful use of a weapon and the advantages derived from such use is what brings and keeps them together.

However, our experience in the Sundarbans suggests that subordinates do not form a homogeneous category. On the one hand, the workers do engage in a variety of subversive acts, yet, on the other, they compete with each other and, if necessary, discredit each other in trying to be in the good books of the contractors, engineers or panchayat leaders during the latters' visits to construction sites. As a result of this, the more successful ones get recruited as supervisors or dafadars, as happened in the case of Dinu or Srinath. Dinu or Srinath's story of how one becomes a dafadar explains why labourers want to be in the good books of the irrigation or panchayat functionaries. As a dafadar, one wields power in supervising one's colleagues. There is also a keen competition between Dinu and Srinath for equitable distribution of supervisory responsibilities. Both are watchful of each other and pay frequent visits to the Gosaba irrigation office so as not to lose out in the competition. In a terrain where people constantly shift their subject positions, practices necessarily alter before they can ossify.

In his more recent and provocative work, *The Art of Not Being Governed: An Anarchist History of Upland Southeast Asia* , Scott (2009) moves away from his earlier framework developed in *Weapons of the Weak* (1985) or *Domination and the Arts of Resistance* (1990) and focuses on communities indulging in state evading strategies and 'opting for statelessness' (Randeria 2010, 464). These communities are without the so-called civilisational history, for according to Scott, 'The relationship of a people, a kinship group, and a community to its history is diagnostic of its relationship to stateness' (Scott 2009, 237). Here we are presented with a region stretching from northeast India, through Bangladesh, Burma, Laos, Cambodia, Thailand to the central highlands of Vietnam. Scott argues that for a long time subaltern communities and 'peripheral and acephalous groups' (Ibid) inhabiting this region adopted a way of life that escaped centralisation and state formation. These groups' nomadic livelihood, de-territorialised existence and incoherent ethnic identities are believed to have posed significant challenges to the system of governance. He refers to hunting and shifting cultivation indulged in by these communities as deliberate techniques of livelihood against wet rice cultivation and the irrigation systems founded by the centralizing state (Randeria 2010, 465). People deliberately adopted the above strategies that required low labour employment and allowed for little surplus generation (Ibid). To quote Scott, '... the condition of remaining "legible" to the state and producing a surplus that is readily appropriable is embedded in the concept of civilization' (Scott 2009, 101). Scott seems to provide a counter point to governmentality predicated upon the principle of legibility, that is, making people's lives transparent and therefore amenable to governance. However, it is these hill men and the incoherence of their tribe and ethnicity that had escaped the modern state's penchant for classificatory order based on cadastral survey and enumeration through census (Ibid, 238).

Scott valorizes neat binaries. His thesis makes the ruler and ruled appear as though they are homogenous categories opposed to each other and thereby glosses over the interconnections that exist between them. Scott's formulation is problematic insofar as he seeks to situate the state repellent strategies of the subaltern outside the ambit of the modern state. The question that seems relevant is whether these strategies are produced in vacuum or through a constant negotiation between the ruler and the ruled. What the subalterns do to resist governance is very often informed and shaped by their responses to the strategies deployed by the governmental power. This chapter has considered the strategies of embankment workers, labour supervisors and those of the displaced islanders in Garantala. These people and their strategies of subversion or protest crystallize only in the context of their negotiations with governmental machinery. It is in response to governmental apathy and neglect that they appear as selves capable of comprehending what their entitlements are and adopting strategies of subversion or protest. To situate these people outside the ambit of the state or to attribute the origin of their activities to their bounded and impervious life-world would be to reify their concrete agential role.

Notes

[1] Chamber's seminal work (1983) *Rural Development: Putting the Last First* opens up a new vista in development thinking and practice. His subsequent works including the one mentioned above induce a paradigm shift in development discourses, putting people in the forefront of development processes. He presents two paradigms as binary opposites problematizing the top-down development model. For details, also see Chambers (1994, 1997).

[2] Our discussion here is influenced by Hill's views on riparian law and its impact on colonial settlement policies in the Gangetic delta particularly in the Kosi diara estate in the Bengal Presidency. I am solely responsible for the arguments I have presented here. I draw on Hill's analyses to historically situate the Sundarbans in the larger discourses of colonial settlement policies in northern India. For details see Hill (1990, 1991).

[3] Considering the devastation that the river caused over centuries, the Kosi is very aptly called the River of Sorrow in Bihar.

[4] An employee who belongs to the lowest rung of the government machinery.

[5] The villagers often use the words together. This suggests that the everyday workings of the party and panchayat make the distinction that formally exists between them appear fuzzy.

[6] When agricultural land gets flooded in saline water because of embankment collapse, owners of the salinated lands often covert their lands into fishery for prawn or fish cultivation. They generally lease out lands to moneyed people from Kolkata who are interested in pisciculture. I will discuss embankment, fishery and prawn trade in detail in chapter 6.

5

Beldars, Embankment and Governance
Question of Aboriginality Revisited

Biru Sardar and the tiger[1]

In the Sundarbans, the so-called land of tides and tigers, people's livelihood revolves around the forest, water and narrow creeks. Stories of human encounters with tigers abound. After a day's work, when people meet at local tea or grocery shops on the islands, tiger stories often figure in their otherwise mundane conversations. The most interesting story that I heard during my fieldwork in Kusumpur was how an adivasi Sardar once killed a tiger. I heard the story for the first time as it was told by a few villagers, who happened also to be the workers of a local voluntary organization (Sangathan),[2] when one evening they congregated at the tea stall next to the organization's office. The workers were sharing tiger stories amongst themselves: stories of tigers killing people and instances when people escaped death at the fangs of the tiger. Suddenly Ratan, a worker of the Sangathan, said, 'But nothing beats the story of Biru Sardar'. 'There you are', everyone present in the stall instantaneously agreed with him. Turning to me, Prafulla, another worker of the Sangathan, said, 'You might like to hear this story, as you seemed interested in knowing about the adivasis in Sardarpara [locality where Sardars live]'. Then Prafulla and the others asked Ratan to narrate the story. Ratan happened to be the first narrator but later I heard the same story from other villagers who had all heard it from their fathers and grandfathers. According to the villagers, the incident happened when Kusumpur had already become inhabited. The story goes like this:

Once, on a warm and sultry night, Biru Sardar, an aged tribal of Sardarpara, was sleeping outside his house. Because of mosquitoes, he had covered himself with a sheet. Biru was in deep sleep when a tiger appeared. As he was fully covered, the tiger could not make out if it was a human being, but continued towards him. Meanwhile Biru's sleep had been broken by the smell of the tiger, but he pretended to be still asleep. The tiger came near Biru and, in an attempt to know what it was, finally had him between its four legs. Realizing that he was lying under the tiger, Biru, in a state of shock, suddenly embraced the tiger, held it hard against him and shouted, "Tiger", "tiger". Hearing Biru screaming, his neighbours came running to his house and found him in that state. Initially, they found it extremely difficult to separate the two. But when they did so eventually, the tiger was found to be dead. Biru had embraced the tiger so hard that it was suffocated to death.

After the story was narrated, one of the villagers asked me, 'Do you think a normal human being could do that?' Bankim, another worker of the organization exclaimed, 'Such strength, my goodness! Only Sardars are capable of that'. 'Why do you think so?' I asked, to which Bankim replied, 'They are the primitive people of our country. They have supernatural strength and the ability to toil in their blood'. After a while, Ratan said, 'Yet, once Sardars become government beldars, they do nothing to maintain the embankment'. 'Why do you say that?' I asked him. Ratan answered, 'You don't need to ask me, you ask anybody, he will say the same thing. If you intend to meet the beldars, my advice is don't meet them at the beginning of the month', Ratan cautioned. 'Why?' I asked. 'That's the time they get their salary and they are busy boozing and blowing away their money'. Bankim intervened, 'Most of the time they remain drunk and do nothing worthwhile. Over these years the condition of the embankments has worsened due to their laxity and negligence'.

The chapter focuses on the adivasi or tribal population of the Sundarbans. The Sundarbans reclamation was done by the tribals indentured from Chotanagpur (Das et al. 1981, 90). These adivasis mostly belong to Santhal and Bhumij tribes whose ancestors were brought to the delta by their colonial masters for reclamation. The tribal people indentured from Chotanagpur Plateau cleared jungles, chased tigers, built embankments and made the wetland habitable. The story of Biru Sardar and the conversations that followed show that Sardars were remembered for their physical strength and ability to toil. They were portrayed as tiger chasers and embankment builders in the Sundarbans. Today, the Santhal and Bhumij Sardars – the descendants of illustrious forefathers – are beldars, the irrigation department's staff entrusted with the task of protecting and maintaining the embankments in the Sundarbans. Their livelihood revolves around the physical landmass encircling the islands, its protection and maintenance. The chapter begins with the story of an adivasi Sardar embracing a man-eating tiger to death. This particular story and others that I subsequently document in the chapter portray adivasis as humans with extraordinary strength and courage. The next section provides a brief outline of the conceptualization of labour and aboriginality that came to operate in colonial labour market and situates the idea of adivasis as 'wild people' in the broader discourses of aboriginality and labour against which I seek to explore the non-adivasis' perceptions of the adivasis in the Sundarbans. I then focus on the narrative of an adivasi bauley (a woodcutter and one who knows mantras to charm tigers in the jungle) which contests the dominant image of the adivasis as tiger chasers or jungle clearers. I highlight the bauley's story to show how important it is to document voices and narratives that seem marginalized by a more hegemonic representation of the adivasis as wild people. I turn then to the question of the adivasis' identity as embankment builders. Here we also come across stories that on the one hand celebrate the adivasis' historic role as reclaimers of the Sundarbans or embankment builders (thereby seeking to establish an unproblematic correspondence between aboriginality and labour power) and on the other, accuse the present day adivasis, who serve as government-appointed beldars, of not being aboriginals any more. In other words, the adivasis as beldars are perceived as having failed to reproduce the ethnic qualities of their forefathers. At the end, I present beldars' narratives from the present day Sundarbans to show how such overarching representation of the tribals as wild people produces only a reified notion of the adivasi self. Here, beldars' own predicaments can be seen as expressive of their right to self-representation.

The colonial labour market and the discourse of aboriginality

The story and the conversations that followed remind us of the centrality of colonial discourses of primitivism and aboriginality. In writing the history of shamanism and conquest in the Amazon basin, Taussig (1984) shows how colonialism constructed the idea of a 'wild man' and unleashed a regime of terror necessary for the spread of rubber plantation in South America.

> Putumayo rubber would be unprofitable were it not for the forced labour of local Indians, principally those called Huitotos. For the twelve years from 1900, the Putumayo output of some 4,000 tons of rubber cost thousands of Indians their

lives. Deaths from torture, disease, and possibly flight had decreased the population of the area by around 30,000 during that time (Taussig 1984, 474).

Colonists in the Amazon constructed the locals as savages. They were struck by the locals' knowledge of the forest and their ability to detect sounds and footmarks where white men perceived nothing. On the trail of an animal, they would suddenly swerve, then change again as if following the scent of their prey (Ibid, 487). Such 'savagery' needed to be contained and channelled in the direction desired by colonial capitalism. The result was the flourishing of rubber plantation along the upper and lower reaches of the Putumayo river in Columbia. A new equation was drawn between primitivism and civilization whereby the idea of a 'savage' was deliberately invoked and deployed to perpetuate the savagery of civilization. To ensure that the extraction of rubber was adequate, punishments were meted out to the Indians. Flogging was the dominant form of punishment. 'Indians were flogged when they brought in insufficient rubber ... Flogging supplemented other tortures, such as near drowning and choking, designed, as Casement points out, to create a space of death'[3] (Taussig 1991, 39). Mothers were often flogged because their little sons had not brought enough rubber. Deliberate starvation was resorted to, at times to frighten, more often to kill, and prisoners were kept in the stocks until they died of hunger (Ibid, 39–40). This was deliberately done to bring civilization to the land of the so-called savage.

However, the happenings on the Putumayo were not an isolated incident in the history of colonialism. The colonial labour market as it came to operate in India also saw the emergence of tribals as coolies and their recruitment as indentured labourers for the expansion of colonial capitalism (Mohapatra 1985; Breman and Daniel 1992; Prakash 1992; Bates and Carter 1994; Ghosh 1999). In this process, the rebellious hill dwellers of Chotanagpur were transformed into 'figures of docile and hardworking coolies' (Ghosh 1999, 14). Furthermore, the colonial market both produced and consumed primitivism and fetishized the correspondence between aboriginality and physical strength as a crucial element in the construction of adivasi personhood. Prakash (1992) makes an interesting observation about the perceived relation between aboriginality and cooliehood in the context of exhibitions on aborigines held in colonial India to institutionalize the pursuit of anthropology as the science of races. Prakash quoted George Campbell as having observed that an exhibition of the aborigines would be the easiest thing in the world because the exhibits were such excellent labourers that they might be utilized as coolies to put in order the exhibition grounds at certain times, while at others they would take their seats in exhibition stalls for the instruction of the public (1992, 158). Campbell persuaded the Asiatic Society to hold such exhibitions where members of different races could be assembled for presentation as living exhibits, suitably framed in classified stalls and could be observed in motion as functioning objects (Ibid). Mohapatra and Ghosh (1985; 1999) argue that aboriginality became a new language for classifying Indians as labour and ethnic stereotypes operated as the basis for recruiting coolies for the flourishing plantation economy in Assam and coal mines in the Chotanagpur Plateau.

A classificatory scheme arose whereby coolies considered to be 'first class' included the Chotanagpur hill people such as the Bhumij, Santhal, Oraon, Munda and Kol, while those ranked below them comprised Khettris from Bihar and other castes from Bengal, Bihar and the North-Western Provinces (Ghosh 1999, 32). It was clear from this classification system that first-class coolies should be 'pure' aborigines or 'primitives' (Ibid).[4] It was because of this construction of aborigines as 'better' coolies that the prices fetched by an aboriginal coolie were higher than those fetched by others (Ibid, 34). Thus, it was the perceived distinctiveness of the Santhals, Bhumijs and Mundas which allowed them an entry in the colonial labour market. Paradoxically, the market, which was supposed to erase cultural differences, actually operated through upholding a distinct ethnic stereotype (Ibid, 35).

Mohapatra argues that labour recruitment to the coal mines in Chotanagpur region followed a particular pattern. The so-called semi-aboriginals and landless castes of Manbhum and Hazaribagh went to coalfields, while the tribals like Oraons, Mundas, Bhumijs, Kharias, Hos and others were recruited for tea gardens of Assam (Mohapatra 1985, 262). It is interesting to note that owners of the coal mines recruited labourers from the 'up country' which included places like Gaya, Monghyr, Saran, Patna and also districts of Bihar and North-Western Province districts of Allahabad, Unnao, Pratapgarh and so on (Ibid). One wonders why the Chotanagpur tribals living much closer to the coal mine belts were sidestepped in search of labourers from the up country. The reason that was found doing rounds in the region was that the aboriginals of Chotanagpur abhorred working underground in the mines (Ibid, 263). However, for Mohapatra, a more significant reason was that the aboriginals were found unsuitable for coal mines. A stereotype of the local adivasi mining population as idlers, living for the day, thriftless, addicted to drink and nomadic began to emerge in the official reports and mine owners' complaints (Ibid). However, the aboriginals were most sought after when it came to recruiting workers for tea gardens in Assam. The aboriginals were transported to the plantations of Assam because they were described as having the expertise to clear jungles and being better able to survive the 'malarious climate of Assam'. Even though the mortality rate in malarious Assam among the Chotanagpur aboriginal coolies was the same as that among the coolies of North-Western Provinces, the stereotype of aboriginal coolies as sturdy, robust and extraordinarily adaptive to adverse working conditions continued to inform recruitment in the colonial labour market.

It was the prospect of revenue generation that prompted colonial authorities to reclaim the Sundarbans wetlands. From the early part of the nineteenth century, the British assumed proprietary right to the Sundarbans and began leasing out tracts of the forests for undertaking the clearing operations preparatory to planting paddy. The colonial power embarked upon the rapacious reclamation of the wetlands, and the adivasis from Chotanagpur were deployed as coolies for clearing jungles and chasing tigers away from their habitat. It was no accident that coolies brought in as indentured labourers for the purpose of Sundarbans reclamation were also members of the Santhal and Bhumij tribes (Pargiter 1934, 57; Das et al. 1981, 90). According to Choudhury and Bhowmik (1986, 330–331), like the indigo plantations in Bengal or tea plantations in Assam, clearing of these large tracts of forest and marshy lands of the southern part of Bengal required cheap labour. Naturally, the most tempting target was the adivasis who were deemed sturdy and hardworking persons able to live at a sub-human level of existence (Ibid, 331). Thus, it is the ability of humans to work at a sub-human level, which led to the privileging of adivasis over their colleagues from North-Western Provinces or the up country. Who could work at a sub-human level except the adivasis stereotyped as wild and nomadic? Adivasis as the 'unsettling coolie' (Breman and Daniel 1992, 279) facilitated the colonial commercial expansion whether such expansion extended to malarious Assam or disease-prone mangrove swamps of the Sundarbans wetlands. The image of the adivasis as wild and nomadic provided colonial rule with an opportunity to transport them anywhere.

Adivasis as tiger chasers, adivasis as savages

Thus, historically, the adivasis played a significant role in the Sundarbans reclamation. And the case of the Sardar suffocating a tiger to death provides an instance of how stories circulating in the locality perpetuate adivasi Sardars' career as forest clearers and tiger chasers. During the early stages of my fieldwork, I heard yet another story illustrative of the predatory strength of the Sardars. I heard this during a rainy evening when along with many others I took shelter in a local club. It was a monsoon rain and realizing that it was not going to stop soon, the villagers who were stranded there were having tea and conversing amongst themselves. This was the time when I came to Kusumpur and was new to the people I met at the club. Therefore, they were generally curious about my research. Their queries triggered off a discussion on the Sundarbans reclamation and its history. Eventually, Sardars figured in our conversations. The villagers narrated how they heard the Sardars had cleared jungles during the day and spent their nights sitting on trees. The story they narrated was as follows:

> It was while sitting on a tree that an adivasi Sardar once dozed and fell to the ground. Before the Sardar regained his senses, he saw a tiger pouncing on him. Seeing the tiger about to grab him the Sardar caught hold of its fore legs. As a result the tiger suddenly lost control of its body and fell to the ground on its back with a heavy thud, its fore legs still firmly held by the Sardar. The tiger lay on the ground and continued to purr as if it was rendered motionless by the grip of the Sardar. The Sardar held on to the tiger for some time before he finally set it free and chased it out of his sight into the jungle.

It is something of a surprise that the villagers, who narrated the above story demonstrative of Sardars' strength and courage, also despised them as savages. In the club that evening, the villagers also discussed with disdain the predatory tastes and habits of the Sardars. According to them, until recently they were savages. Even a few years back, it was difficult to pass through Sardarpara. The place used to be smelly and dirty. It was only now that they had acquired a semblance of civility. Sensing that I might have been unconvinced by their reflections on the Sardars, the villagers drew my attention to a person sitting rather quietly in the club room saying, 'Ask Kamal to narrate his experiences at a Sardar's house in Sardarpara'. Kamal was a teacher at a primary school in Kusumpur, and he, I could make out, had something to share in support of what was being discussed. I requested Kamal to share his experience, and finally with the prompting of others present in the club room, he narrated his experience when he was invited to Kanak Sardar's house.

> Kanak Sardar's son was Kamal's student at Kusumpur school. As a mark of respect for his son's teacher, Kanak Sardar once invited Kamal to his house for

lunch. He was given a whole catla mach [Catla catla, Indian carp] to eat. Kamal
was puzzled as he had no clue as to how to tackle the fish. Looking at the fish
Kamal felt nauseous because it looked as though it was barely cooked and not
properly de-scaled. But seeing Kanak's family members around him and the
eagerness with which they wanted to feed their guest, Kamal proceeded to
do justice to the food served. As he bit into the fish, blood oozed out of its
head. The smell of blood made him feel sick and he could not eat any further.
Kamal suffered from nausea and loss of appetite for about a month. Since then
no one had ever accepted an invitation to a Sardar's house.

'We all eat fish, but can you eat fish like that? Can you think of serving fish
to your guests in that manner?' Kamal shook his head rather disapprovingly.
The stories narrated above construct adivasis as an exclusive category whose
ways of life were not comparable to those of non-adivasis. It was as if their
strength and courage lay in their savagery, in their being wild people.

The voluntary organization, Sangathan, run by the local people, which
I have mentioned at the beginning of the chapter, was one of the many
organizations implementing social welfare programmes in the Sundarbans
and South 24 Parganas. It implemented various central-government-funded
programmes including housing for the poor particularly those belonging to
scheduled castes and tribes. While I was in Kusumpur, I was witness to a
controversy concerning the government-funded housing project in Sardarpara
of Jagatpur village in Kusumpur. During my stay in Kusumpur, I was an
occasional visitor to the organization's office. It was during one such visit to
the Sangathan's office that I found some of the workers engrossed in heated
conversations with the coordinator of the organization. The workers did
not seem to be in the mood to listen to the coordinator even though he was
trying to explain things to them. Ramen, the coordinator, insisted that his
colleagues should go to Sardarpara to solve a problem that had cropped up
in connection with the organization's housing project. However, I found that
Ramen's colleagues were not prepared to listen to him and visit Sardarpara.
They categorically said that they would not spend time in the company of the
uncivilized people who had not even set up individual toilets and still defecated
in the open. According to the workers, there was an awful smell of night soil
pervading the whole of Sardarpara and there were flies all over the place.

Later when I did ask Ramen about the problem in Sardarpara, he said that
a widow in Jagatpur had been allotted a house under the central government's
housing scheme. Recently the widow died and since she did not have an heir, the
young generation of Sardarpara wanted to convert her house into a local club.

Ramen told me that he did subscribe to his colleagues' perception about Sardars but still he wanted the workers to meet the Sardars as part of their official duty and settle the dispute. Here we are not concerned with whether the workers finally went to Sardarpara or Ramen successfully handled the problem. The workers' reluctance to visit the Sardars and their comments are demonstrative of their perceptions about the adivasis. On the one hand, they celebrated the adivasis' strength and courage, and on the other despised them as savages. It is interesting to note that a voluntary agency that claimed to represent the interests of the Sundarbans islanders and even implemented programmes for the tribals did not have any member of the adivasi community within its organizational fold. Despite being an intrinsic part of the social map of the Sundarbans, in the eyes of their neighbours, adivasis remain 'wild' people, 'yet to be civilized'.

The non-adivasi perception of the adivasi food habit or dietary preferences as revealed in the above story has its genesis in the way colonial rule made distinction between castes and tribes. According to Skaria, the distinction between castes and tribes drew on and was made possible by colonial constructions of wildness (Skaria 1997, 727). Skaria shows how the grounds for the distinction between castes and tribes was already being laid by the late eighteenth century. He particularly drew attention to William Jones's theory about an Aryan invasion in India. An idea gained ground that tribals or aboriginals were the original inhabitants of India, while upper castes were the descendants of Aryan invaders (Ibid, 729). Skaria has quoted John Briggs[5] as having provided a list of differences between aborigines and Hindus in 1852. These differences are as follows:

> Hindus had caste division, aborigines did not; Hindus did not eat beef, aborigines did; Hindu widows did not remarry, aborigine widows did; Hindus abhorred the spilling of blood, aborigines reveled in it; Hindus ate food prepared by their own castes, aborigines ate food prepared by anyone; and so on (Ibid).

With the passage of time, arguments advanced to distinguish tribes from castes became much purveyed common sense (Ibid) informing stereotyping of an adivasi self. It is in the telling of such stories that ideology and ideas become emotionally powerful and enter into active social circulation and meaningful existence (Taussig 1984, 494).

The adivasi perception of themselves as vulnerable

As one enters the adivasi world via these stories, one feels as if Sardars in the Sundarbans stand out as tiger fighters or chasers. However, my meeting with

Mangal Sardar was significant in many respects. Mangal was a frequent visitor to the forest because like many other Sundarbans islanders, he 'does jungle' (jongol kore).[6] I remember my first meeting with Mangal at the Sajnekhali Forest Office where I had gone to meet the ranger. As I entered the office premises, I found Mangal talking to a forest guard who introduced me to him. From their conversation, I had gathered that Mangal was an occasional visitor to the forest office and today he had dropped by on his way back from Gosaba. Mangal took a bidi from the forest guard and the following conversation took place:

Mangal: How long do you take to issue honey collecting passes [permit for honey collection] these days?

Guard: I am not sure how long it will take, but to avoid delay, put in your request in advance.

Mangal (impatiently): Why don't you people hasten the process of issuing passes?

Guard: Who am I to hasten the process? Everything has to be in accordance with the rules.

Mangal (to me): These people are becoming unnecessarily bureaucratic, imposing stricter forest rules which make it difficult for us to survive.

Guard: If the forest office does not make rules, then you people will be in trouble; you will meet with accidents in the forest.[7]

Mangal (pointing to his own stomach): When you are hungry and your hunger drives you crazy, you forget all rules, you enter the forest and meet your destiny.

My second meeting with Mangal was at his house in Kusumpur when he came back from catching crabs in the forest. While Mangal was drying his fishing net in the sun and explaining how they used their curved iron rods to drag crabs out of their holes in the soil, a tiny little mud hut with an idol of Bonbibi[8] inside located in one corner of the courtyard suddenly caught my attention. I asked, 'Do you worship Bonbibi as well?' 'Why not?' Mangal looked a bit surprised. 'All of us who work in the forest worship Ma Bonbibi'. 'But I thought only those who feared the tiger worshipped Bonbibi? Your forefathers I heard were valiant tiger chasers who fought tigers and reclaimed the Sundarbans. As legitimate successors, you, I thought, could do without Bonbibi'. Mangal smiled and answered, 'Living in this land of tides makes you respect the rules that Bonbibi has set for those entering the forest. The forest does not belong to any individual; all that grows in the forest ought to be shared by all living beings. Unless you learn this basic principle by heart and enter the forest with a pure heart (pabitra mone) and without a sense of greed, you cannot earn your living from the jungle'.

I further asked Mangal about how they protected themselves against the tiger. According to Mangal, before they entered the forest they placed their hands on the soil and chanted mantras to sense if the tiger was around. If their hands gently settled on the ground, they felt that it was safe to enter the forest. But if their fingers started to quiver, they left that part of the forest and moved elsewhere. I was surprised to know that Mangal was both a tiger charmer and a bauley who led his team into the forest. Mangal further stated that there were different mantras – some meant to lock the jaws of the tiger so that he cannot open his mouth, others to make the tiger change his path. He also cautioned that mantras to charm tigers should not be used indiscriminately, but should be used only when danger was imminent. And it was absolutely imperative that before leaving the forest, a bauley should chant the same mantras to free the tiger and other animals from his spell so that they could again move freely in the forest. I interrupted Mangal while he was reciting a few lines from Bonbibi's Johuranama[9] and asked if he ever encountered a situation when his mantras did not work. Mangal suddenly became quiet and said that despite his knowledge of the mantras, he could not save his friend and colleague. I could sense heaviness in his voice as he uttered the last few words. He was hesitant to tell me how his friend was killed in the jungle, but at my insistence, he finally narrated the incident which happened some years back during the honey-collecting season:

> Mangal and seven others went to the forest in search of honey. The fortnight-long expedition to the forest had nearly come to its end. They had collected honey enough for the entire group. Before they finally left the forest, Mangal, being the group leader offered puja to Ma Bonbibi, saying that they got enough and were satisfied with what Ma Bonbibi had given them. As they prepared to board the boat, Kesto Sardar, a member of the group and its sajuni [one who prepares and readies the boat for the forest expedition], suddenly decided to go back to the forest in search of some more honeycombs which he had spotted on his way back. His decision to go back came as a surprise to the other members of the group because one is not supposed to re-enter the forest after one has finished offering puja to Ma Bonbibi. Despite their persuasions, Kesto was adamant. This posed a dilemma for Mangal: on principle, he was against re-entering the forest and at the same time he could not have left a group member behind especially when the member was his sajuni. When

he found that Kesto was adamant, he reluctantly followed him back into the forest. As they entered the forest and stood under a relatively big honeycomb which Kesto was preparing to break, Mangal heard something move in the bush behind them. Before Mangal could turn around to see what it was, he saw a tiger jump past him, grabbing Kesto by the neck and disappearing into the forest. The whole thing happened in a split second and when Mangal realized what had happened, he shouted to his colleagues for help. Not being able to locate which way the tiger had gone, they started to let off crackers to frighten him away from his food. They kept doing that until they managed to find Kesto's body in a pool of blood with one arm and genitals gone.

In trying to recollect the incident, Mangal looked horrified. According to him, the tiger attacked with a lightning speed, a speed that one could experience but would always be mortally scared to recollect. 'But then didn't you use your mantras?' I asked. Mangal smiled and answered, 'When you are driven by greed and lust, nothing seems to work. Even now I feel sad when I am reminded of Kesto's death. But then it was Kesto's greed which misled him. Ma Bonbibi would not protect anyone who wants more than he needs'. Mangal's smile makes his story more poignant. As a bauley, Mangal has mantras at his fingertips, but his approach to using them reflects his deep sense of responsibility and commitment towards his profession as a jungle doer. For the Sundarbans islanders like Mangal, the forest is the realm where egalitarian principles are at work making humans and non-humans respect each other's claims.

However, the most significant aspect of Mangal's narrative is that it runs contrary to the dominant representation of adivasis as jungle clearers and tiger chasers, as people of predatory strength and courage. It was as if by being a victim of the tiger that Kesto proved to be an ordinary human driven by the lure of a livelihood in the forest. Kesto Sardar's death and Mangal's predicament was the moment when the adivasis could throw the ancestral burden off their shoulders and appeared as selves freed of the shackles of colonial servitude. Thus, the haunting tiger ceases to haunt any longer, the predators (the tiger and the adivasi) stand face to face with the masks fallen off their faces.

From embankment builders to beldars: Perceptions of today's adivasis as cutting a sorry figure for their illustrious forefathers

Another arena where non-adivasis' perceptions of the adivasis assume a significant form is in embankment building and protection. The Bhumij and Santhal adivasis who cleared the forests and reclaimed wetlands were engaged in embankment building activities as well. After the abolition of zamindari rule in the 1950s, when protecting and maintaining the Sundarbans embankments came under the purview of the irrigation department, the department recruited the adivasis to the post of beldars, government employees responsible for protecting and maintaining embankments. The rationale behind such recruitment was not only to employ local resources to maintain embankments, but also to pursue a wider government policy of uplifting 'socially backward' communities. The adivasis, with their assumed aboriginality, physical prowess and historically defined role in the Sundarbans reclamation, qualified for the label 'indigenous people', an oft-quoted phrase in contemporary government discourses of development.

Adivasi Sardars' role as embankment builders is as much an object of gossip among the villagers as is their role as forest clearers and tiger chasers. The only difference here is that the narratives that one comes across are more accusatory in nature. Soil erosion and embankment collapse spell disaster for everyone, and their frequent occurrence is often attributed to laxity and negligence on the part of the beldars. While I was in Kusumpur, embankments collapsed on a number of occasions. During one such major embankment collapse in the southern part of the island, when I reached the site, I found that people from nearby settlements had already started urgent repair work. The villagers who lived close to the site were digging earth from the paddy fields in an attempt to create a new stretch of embankment behind the one that was on the verge of collapse. The stretch of embankment that had caved in already looked extremely vulnerable. A huge chunk of earth was leaning towards the river. Drawing my attention to it a villager said, 'All this is due to the negligence on the part of the beldars: they do not bother to visit these spots, neither do they add earth to embankments'. 'Is it the duty of the beldars to add earth to these mounds?' I asked. 'Who else?' he asked in return. The man then shouted out to a teacher of a local primary school, 'Sir[10], tell this gentleman what the beldars do these days'. The man who was inspecting the breaches that appeared in the embankment said, 'What can I say about the beldars, the less said the better. They have become sarkari baboos [government clerks] who enjoy a nice salary and do nothing to protect embankments'. 'Why do you say that?' I asked. In response to my question the teacher said, 'Look at their activities and compare them with what their forefathers did. Their predecessors were embankment builders. From my grandfather, I heard how Sardars worked tirelessly throughout the day and had only saline water to drink. Sardars had extraordinary physique and enormous courage, otherwise why would they be employed for jungle clearing and embankment building?' 'Was it because of the nature of their work that they were dark complexioned?' The villager who had introduced me to the teacher intervened, to which the teacher responded, 'No, it is because of their dark complexion that they can toil so much. No matter how long they work in the sun nothing will happen to them'. Turning to me the teacher said, 'My grandfather told me a story of how a Sardar once stopped water seepage through a hidden hole in the embankment. During high tide when the water level is sufficiently high, water seepage takes place through these holes. If they are not plugged in time, saline water can contaminate the soil and rice fields. I heard this story from my grandfather, so it must have happened long back':

> Once during high tide, it was discovered that water seepage was taking place through a ghog [hidden hole] in the northern part of Kusumpur. It was the beginning of the high tide. Water was on the rise and the hole detected was found to be gradually enlarging. Realizing that something needed to be done urgently, a middle-aged Sardar, called Charan or Ramu, suddenly stood with his back to the embankment wall, thinking that his body could serve as a shield to prevent the water seepage. He stood there for six or seven hours pressing himself hard against the mud wall. When water receded during low tide, it was found that saline water had eaten into his skin and left deep holes on his body.

The teacher added, 'Such was the strength and commitment of the Sardars, but look at their successors and how they are wasting their bodily strength. They have become idle and their only interest is in their monthly salary. At the beginning of every month you will find them boozing and blowing away their money'.

Not only are Sardars perceived as having failed as embankment builders, but they are even considered to have failed in reproducing their ethnic stereotypes. Here, it is interesting to observe further twists in the language of aboriginality and labour. Colonial labour market encountered problems of drinking and idling among the aborigines. The adivasis, who were the planters' favourite because of their expertise in clearing jungles and their ability to adjust to plantation life, were not considered suitable for coal mines where the need for mining coal throughout the year required the labour to be more disciplined (Mohapatra 1985, 265). However, what we hear today in the Sundarbans suggests that drinking and idling, which once characterized the adivasis and served as significant ingredients in constructing their ethnic stereotype, are now seen by the non-adivasis as symptomatic of tribals losing their ethnic stereotype and becoming estranged from their 'tribe-hood'. The teacher's narrative seems to suggest that by becoming idle and alcoholic the adivasis fail to live up to the idealized image of the tribal as labourer with uncanny strength and courage.

Letting the burden off their shoulders: Embankments and predicaments of beldars

Because of these accusations, I wanted to meet the beldars of Sardarpara. So I made contact with them and informed them of my intention to visit them. And it was decided that they would meet me at Bimal Sardar's house at Sardarpara in Kusumpur. The day I went to Bimal's house I found that they were discussing something amongst themselves. Seeing me Babulal, one of the beldars, asked me, 'How long do you think you are going to take?' 'That depends on the time that you think you can spare', I answered. Babulal gave an awkward smile and said, 'Actually a problem has cropped up'. Then pointing to a young man sitting at a distance he told me, 'His father was a beldar who died recently. Since he died while in service, his son now has a legitimate claim to his father's job. But the irrigation department is creating a problem on technical grounds. We have decided to take this boy to the Gosaba office and sort it out. Therefore, you need to let us off in an hour and a half'. The boy, who was sitting with a stick in his hand and clad in two pieces of cloth,[11] one to cover the lower part of his body and another piece wrapped around his bare chest, looked to me like someone whose father's formal funeral rites were not yet over. I told Babulal, 'An hour and a half is okay, but it seems that his father's death rituals are not yet over'. Babulal replied, 'That's precisely the reason why we want to take him now to the office and convince the officials about his bona fide claim'.

With an hour and a half at my disposal, I started conversing with the six beldars. Except for Karna Sardar, the other five beldars had all inherited their fathers' occupation. Initially, our discussion centred on their fathers and forefathers and their deeds during the time of reclamation of the Sundarbans. They described the odds they faced in clearing forests during the early stages of settlement. While we were chatting, Bimal Sardar kept drinking rice alcohol and nodding his head as a mark of his tacit support to what his colleagues were saying. He was drinking and at times passing his tumbler around to others for a sip. As soon as his tumbler was empty, his wife came to replenish it.

However, the moment I asked them about the allegations against them Bimal turned grave. Suddenly he got up as if the question shook him out of his inertia and asked, 'Who told you this? Do these people know what our job and duties are?' He did not bother to wait for my response and continued, 'We are not irrigation labourers, meant to add earth or repair embankments. Our job is to detect ghog in the embankment. Detecting these ghogs is not everybody's cup of tea, you need to have trained eyes and ears to detect them'. Pointing to the distant field dotted with houses and the horizon beyond, he said, 'These people [the villagers] think that anybody can become a beldar'. Bimal paused a while, as he was short of breath.

Babulal went up to pacify him and made him sit on the ground. He then turned to Karna and asked him to explain things to me. Karna said that such holes are made by crabs that come from the sea during high tide. The saline water often flows through these holes and contaminates paddy fields or freshwater ponds. 'You must have seen us move around with long sticks?' Karna asked me. 'No, I haven't', I answered. 'Well, we use these sticks to inspect if there are any such holes. It is only during high tide that we can inspect such holes because it is when the water rises that you can make out if seepage is taking place or not. Sometimes your eyes fail to locate them. Then you need to kneel down and use your ears to make out from the nature and sound of water flow whether holes have developed in the embankments. Holes can be detected only during high tide, but must be plugged when the water recedes'. Our discussion could not proceed further because Babulal and others wanted to go to the irrigation office.

The next day I went to Upen Sardar's house, a few meters away from Bimal's house in Sardarpara. Upen had also been present at Bimal's house when I had met the beldars there. Upen sat at Bimal's doorstep with a crutch by his side which clearly suggested that he was not physically fit. He could not sit through the entire discussion, and before leaving Bimal's place he invited me to come to his house.

Upen's life was in shambles when I met him. He was slowly recovering from a stroke which had paralysed his right limbs, and as he could not work and was fearful that he would not be able to do so any more, he decided to stake his son's claim to his job. According to Upen, the irrigation office in Gosaba had set its face against Upen's claim on the ground that his son was still a minor. Initially, he could not make much headway and his repeated visits to the Gosaba office did not produce the desired results but recently he had approached a panchayat member of his village, who had contacts in the Gosaba irrigation office, to pursue his son's case. We had the following conversation:

> Upen: Now the wind seems to be blowing in my favour. I must ensure that my son gets the job before Kanu's son, the boy you saw at Bimal's place yesterday, gets it. But if the officers decide to recruit him first, who knows they might once again create problems in taking my son in.
> Amites: But don't you think Kanu Sardar's son should get his father's job?
> Upen: It's not that I do not support Kanu's son's claim and I can even be of help to him if need be. But all these things can happen only when I feel secure. I need to look after my own interests as well. For the last five months, my salary has stopped and I do not have much land to my credit. My physical condition is such that working in the fields is out of question. The villagers allege that beldars do not work and it is because of their negligence that the embankment collapses, but did any of them bother to enquire how I am making ends meet?

Beldars' narratives are significant in a number of ways. No matter how the identity of the beldars is construed, either as descendants of their illustrious forefathers known for their legendary physique and strength or as descendants who are no longer capable of reproducing their ethnic stereotype, Sardars as beldars know exactly who they are answerable to. As agents of the state, the beldars see themselves as protecting its interests. Not only did Bimal make it clear that he remained answerable to the irrigation department but the other beldars present there agreed with him. However, Bimal's outbursts also suggest that the beldars, while protecting their interests and those of the state, have also internalized the expectations that people have of a beldar. Identity is as much a product of self-perception as it is the construction of a category by those who do not belong to it (Breman and Daniel 1992, 268).

The way they decided that they would pursue the case of the son of their deceased colleague and establish his claim to his father's job indicated that their role as agents of the state did not make them oblivious of their own rights and privileges. Their lack of familiarity with the detailed rules and regulations of the irrigation office did not prove to be a handicap. Their determination to present the deceased beldar's son in person to the officials of the irrigation department proves that they were ready to counter any such technical problems with the strategies they had at their disposal. Similarly, Upen was on the verge of losing his job, but this sense of loss did not condemn him to a state of utter despair. Rather, it imbued him with a sense of agency that was reflected in his determination to stake his son's claim and pursue it against all odds.

Indigeneity and community revisited

Beldars and their accounts lead us to revisit two concepts that have remained central to theorizing in social science, namely indigeneity and community[12]. I will explain the relation between the two a little later. I first turn to the idea of community. The term community has a very strong emotive appeal especially when it is understood as small, homogeneous, harmonious, territorially bound, ascriptive units where people enjoy face-to-face interactions (Jeffery and Sundar 1999, 37). However, in the words of Chatterjee, the idea of community that seems to have earned acceptance by the industrially advanced western democracies has been that of a modern nation based on the notion of citizenship (Chatterjee 2004, 32). The debate between liberal individualists and communitarians that continued for some time in the Anglo-American political philosophy finally stood resolved in favour of the idea of nation as an inherited network of social attachments (Chatterjee 1998, 277–278). However, this inherited network of social attachments is no longer based on ascriptive, primordial or parochial loyalties because they seemed inconsistent with individualist aspirations of industrial capitalism. Nation becomes an imagined community (Anderson 1991 [1983]). The question of how each of these communities would be fashioned is left to the imaginings of those participating in the building of a nation.

With development becoming a buzz word and the nation-state being the prescriptive format in which political modernization would unfold everywhere else in the world, ascriptive, primordial and parochial loyalties became obstacles to the spread and flourish of the nation in the so-called developing world. In 1951, when the United Nations (UN) declared its policy for the economic development of the underdeveloped countries, it set a high premium on rapid economic progress, declaring that,

> Ancient philosophies have to be scrapped; old social institutions have to disintegrate; bonds of caste, creed and race have to be burst; and large numbers of persons who cannot keep up with progress have to have their expectations of a comfortable life frustrated (United Nations 1951, 15).

Indigeneity or indigenousness became a term in which are couched the primordial or ascriptive bondings of caste, race or religion. Paradoxically, indigeneity was both a characteristic of the developing world and an obstacle to a more progressive community formation in the form of a nation. In other words, the developing world (or the so-called Third World) on the one hand, was looked upon as being replete with exotic indigenous community ties, and on the other hand, it was believed that these ties would progressively disappear with the onslaught of modern institutions and practices.

Interestingly, several years later in an earth summit at Rio de Janeiro in 1992, the same UN, which once castigated indigenous community ties for their being obstacles to growth, upheld indigenous communities as pockets of wisdom and growth and stated that their role in sustainable development was of crucial importance. Agrawal comments:

> The ghost of traditional community had only hovered over Comte's sociology; it has descended to occupy centre stage in current writings on development, environment conservation and resource management ... The current valorisation of community contrasts with earlier analyses that positioned modernity and community at opposed poles. No longer is community the refuge within which tradition lurks to trip progressive social trends. Instead it has become the focus of writings on devolution of power, meaningful participation, and cultural autonomy (Agrawal 1999, 92–93).

Modernization theories during their heyday considered communities of any kind to be obstacles to the growth and spread of institutions and practices of modernity in developing countries. Today 'community' has been brought back onto the global agenda and currently valorized as the only way to achieving sustainable progress. While the development paradigm of earlier years celebrated communities' disappearance, the contemporary environmentalist paradigm only celebrated their return into the development arena. But both provide a very reified view of community. The current valorization of community amounts to reifying the concept in exactly the same way as the teleological theories of progress of the earlier years did when they viewed it merely as an obstacle (Mukhopadhyay 2010, 327). It remains a reified tool in the hands of the policymakers and development practitioners. Protagonists who look upon the 'Third World' as a site for building bounded and impervious community relations fail to acknowledge that concepts such as community, society or indigeneity are constantly being shaped by people's encounter with the modern world. It is therefore necessary to make these encounters and experiences a part of the task of interrogating and redefining such concepts (Chatterjee 1998, 279). Not only are communities, including tribal ones, often hierarchical and conflict-ridden rather than homogenous and harmonious, but also individuals are caught up in overlapping circles of relationships (Jeffery and Sundar 1999, 37). Agrawal argues that community as an organic, harmonious and self-producing unit is problematic because community has two distinct meanings, that is, community-as-shared-understandings and community-as-social-organization (Agrawal 1999, 101). They are not necessarily coterminous with each other. Sometimes it is possible to activate community-as-social-organization, as has often been found in the case of NGOs empowering local-level community networks. But this does not ensure the ready availability of the other dimension of community-as-shared-understandings. It is not easy to establish a direct connection between community-as-shared-understandings and community-as-social-organization (Ibid), for these two meanings do not easily converge.

In this connection, I find Chatterjee's (1998, 2004) argument quite persuasive. Chatterjee argues that in the course of the twentieth century, ideas of participatory citizenship that were so much part of the enlightenment notion of politics have fast retreated in the face of the triumphant advance of governmental technologies that seek to provide welfare to people (Chatterjee 2004, 34). It is no longer the citizen as an ethical category participating in the business of the state. Rather, it is the idea of population that makes available to the government functionaries a set of rationally manipulable instruments for reaching large sections of inhabitants of a country as targets of their policies (Ibid). The policies range from administrative, economic and legal to health, population control, disease, crime and so on. Thus, every conceivable aspect of population life is made available for policymaking governance and control. A certain 'govermentalization of the state', according to Chatterjee, is what characterizes politics in recent times (Ibid). This is as much a feature of the developed world as it is of the developing societies. As a case in point, Chatterjee draws attention to the rise in security policies in the United States or the United Kingdom in the wake of terrorist attacks in these countries (Ibid, 35). In an attempt to provide security and welfare to a larger population, these policies subject people to increasing surveillance and control. Chatterjee's argument has wider implications for the way the term community has been conceived especially in its application to the 'underdeveloped' or developing societies of the East.

In using his formulation to understand the political process as it has unfolded in societies like India, Chatterjee argues that it is the regime of policies that brings the population into certain relationship with the state. However, this relationship does not always conform to strict principles of constitutional politics whereby people emerge as right-bearing citizens (Ibid, 38). Instead, there are varying degrees of negotiations and contestations between those in governance and the governed. The governed do not necessarily appear as though they are armed with knowledge of the constitution or the law of the land, but their constant engagement with the technologies of governance or policies of welfare is what makes them familiar with their entitlements, enabling them to put forth their demands. In such a climate, the so-called indigenous, homogeneous or primordial communities are increasingly found wedded to the technologies of governance. These communities comprise of individuals whose biographies are constantly shaped by multiple policies of the government. Individuals belonging to such communities are neither abstract individual selves nor manipulable objects of governmental policy, but concrete selves acting within multiple networks of collective obligations and solidarities to work out strategies of coping with, resisting or using to their advantage the vast array of technologies of power deployed by the modern state (Chatterjee 1998, 282). This is not to suggest that the communities are no longer homogeneous or shaped by indigenous or primordial ties. But the factors accounting for their homogeneity or primordiality are no longer internal to these communities. Their internal ties are constantly being shaped by a host of agencies or institutions, government or non-government, working outside them. In other words, here we are not concerned with the question of how indigenous the indigenous community identities are. Rather, the question that seems more relevant is how often indigeneity or primordiality comes in handy as an instrument enabling the communities to articulate their entitlements or to grab the resources that go with the governmental policies or programmes. Chatterjee's argument is significant in that it instructs us to look into the career of the concept in a specific context. By providing the instance of a settlement colony along the railway tracks on the outskirts of Calcutta and the settlers' negotiations with various statist and non-statist agencies at various levels, Chatterjee offers an insight into the dynamics of community formation and contests the trend towards reifying the concept (Chatterjee 2004, 53–64). Vasavada et al. (1999) in their research on joint forest protection management in Madhya Pradesh and Orissa have made an observation that bears resemblance to Chatterjee's argument. One of the chief objectives of government's joint forest management in India is to activate local communities that live around forest and involve them in the forest conservation programme. To this end, government creates many forest protection committees. Vasavada et al. argue that to the villagers or local communities, these committees are a means of getting more and more benefits from the government. They see it as an opportunity to get more employment and ensure village development, no matter who does the work (Vasavada et al. 1999, 177).

When we turn to the adivasi community in the light of the above discussion, we find that there are various cross-cutting ties or interests at work within the apparently impervious or homogeneous community of adivasi beldars. These cross-cutting ties and divergent interests make it difficult to conceive of adivasi as community-as-shared-understandings even when we have community-as-social-organization available in the form of an adivasi locality or neighbourhood. However, what keeps the community of beldars together today is not the idea that they are the descendants of adivasis who were once embankment builders and were known for their legendary physical prowess, rather the possibility of their using to their advantage the resources that go with the office of the beldar. This is not to suggest that Sardars as beldars are unaware of their forefathers or their role in the Sundarbans reclamation. However, this awareness of their being descendants of the illustrious adivasi Sardars is relevant to the present day Sardars insofar as it allows them to stake their claim to the government jobs reserved for them. Their determination to collectively pursue the case of the son of their deceased colleague demonstrates their zeal to invoke blood ties as an instrument of negotiation with the government. In this respect, their lack of knowledge about the rules and procedures is not a major handicap. They have various strategies of negotiation at their disposal. Whether Bimal and Babulal pursue the case of the son of their deceased colleague collectively or Upen pursues his son's case individually, they all belong to this community of beldars not simply as descendants of their forefathers, but as individuals who are deeply aware of their predicaments and entitlements as beldars.

Notes

[1]This chapter is a revised version of an article titled, 'Haunting Tiger, Hugging Ancestors: Constructions of Adivasi Personhood in the Sundarbans', which appeared in the Nehru Memorial Museum and Library's (NMML) Occasional Paper Series no. 4 in 2013. Author is grateful to Director Prof. Mahesh Rangarajan and Deputy Director Dr. Balakrishnan of NMML for their permission to incorporate parts of the article in this chapter. I am also thankful to Prof Maitrayee Chaudhuri for allowing me to incorporate in this chapter parts of my article titled, 'Rethinking Community: In the Discipline and Within Development Practices', which appeared in Chaudhuri's edited volume *Sociology in India: Intellectual and Institutional Practices*, published by New Delhi Rawat Publications in 2010.

[2]To respect the confidentiality of its members, the organization has been given a fictitious name.

[3]Taussig here refers to Roger Casement's (an Irish who worked as a British consul in Africa and South America and witnessed British imperialism in both places) report on

colonial rubber plantation along the Putumayo River in Columbia and the torture and violence associated with it. According to Taussig, creating a space of death in Casement's report refers to an elaborate design whereby the oppressor stops short of taking life while inspiring the acute mental fear and inflicting much of the physical agony of death (1991, 39). Taussig's book also establishes a complex connection between terror and healing. It is concerned with both terror and healing, as they have been mediated through narration. Words conveying terror are more terrorizing than terror itself. For details, see Taussig (1991).

[4] This classification was found in E. N. Baker's letter to the Chief Commissioner of Chotanagpur (in *Report of the Commission on the Labour Districts Emigration Act 1880, 253*). I have mentioned the classification as used by Kaushik Ghosh in his essay on indentured labour in tea plantation in colonial Assam. For a detailed discussion of this issue, see Kaushik Ghosh (1999), pp. 8–48.

[5] I have used Briggs' classification as it is found quoted in Skaria. For more details see Skaria (1997).

[6] Doing jungle here refers to activities such as woodcutting, honey collecting and so on.

[7] In the Sundarbans, getting killed by a tiger is very often referred to as an 'accident'.

[8] Bonbibi is the goddess of the jungle in the Sundarbans. She is seen as the protector of those entering the forest with a pure heart and without any sense of greed.

[9] Bonbibi Johuranama (the Miracles of Bonbibi) is a narrative of Bonbibi's win over the tiger demon Dakshin Roy. The text is read out in honour of Bonbibi during Bonbibi puja. The text looks more like prose but reads like verse. The pages of the book open to the right as in Arabic and read from back to front.

[10] This is how teachers are addressed. Sometimes Sir is used as a suffix to the first name of a person who is by profession a teacher.

[11] Two pieces of cloth refers to a ritual cloth that one dons to mourn a parent's death. The cloth is a symbol of austerity for the period between one's parent's death and the formal rites of passage of the deceased.

[12] This particular section is inspired by the ideas of Chatterjee (1998, 2004) and Agrawal (1999). However, I am solely responsible for my arguments and shortcomings.

6

Catching Prawn, Endangering Embankments
Sustainability-Unsustainability Rhetoric

Tiger prawn seed and Baishakhi as a catcher

Baishakhi Sardar, a resident of Jagatpur, lost her right arm three years back. She was catching tiger prawn seeds[1] (locally called bagda/bagda min) in waist-deep water when a shark or kamot attacked her. It caught hold of her right arm and dragged her into deep water. Baishakhi's left hand firmly held on to the fishing net and in desperation she kept hitting the kamot with the wooden frame of the net. Sensing the danger, those fishing with Baishakhi rushed to her rescue. They grabbed her by the waist and tried hard to get her out of the grip of the kamot. Finally, they succeeded, but at the cost of her right arm. Baishakhi was unconscious as she bled profusely. She was in hospital for months before she came back home. While narrating the incident during my visit to her house in Jagatpur, Baishakhi confessed that she still went to the river to catch tiger prawns.

The Sundarbans is a land where fishing folk coexist with those who go to the forest in search of wood or honey. With rivers flowing around the settled and forested islands, fishing becomes a very important activity for the people. People set out on boats to fish at the confluence of rivers and the Bay of Bengal. They also catch fish or crab in the narrow creeks inside the forest. At times, two to three boats are anchored in the middle of the river and nets are cast wide with one end tied to the boats. Fishermen often spend hours and days catching fish. This chapter focuses on the fishers, but not on the so-called traditional fishing communities. Rather, I focus on the women tiger prawn seed catchers such as Baishakhi and people who are involved in prawn collecting, farming and trading in the Sundarbans. It is against the background of raging debates about the environmental sustainability of prawn farming, particularly in the coastal regions, that the chapter delves into the dynamics of prawn trade. However, the chapter does not look into a struggle between the state and fishing community as has been done by Jaladas (2013) in his research on the eviction of fishermen from the Jambudwip island on grounds of violating the state forest act. The second chapter of the book juxtaposes the life of the subalterns against the grand vision of conservation of the Sundarbans' nature. I have shown how people's livelihoods need to be kept at bay keeping in mind the larger issue of conservation. Therefore, the eviction of the fishermen from Jambudwip by the state, as shown in Jaladas' account, is yet another instance of how this conservation works to the detriment of life and livelihood of people.

The chapter intends to go beyond the sustainability-unsustainability debate and unravels different layers of prawn trade and the players involved. The chapter starts by looking at the activities of women prawn seed catchers and their predicaments, particularly in the light of the state's perception of their livelihood as being ecologically destructive. In the eyes of the state, women prawn seed catchers are believed to have depleted the biodiversity reserve and weakened the protective embankments. It is interesting to note that these women are part of a larger network of collectors, dealers and traders whose activities draw us into the domain of prawn politics[2] in the Sundarbans - the politics that is manifested in embankment breaking, fishery making and cross-border trading. Prawn fishery remains the locus around which party rivalries intensify in the region. In the end the chapter shows how the state is itself complicit in the same prawn trade which it labels unsustainable on environmental grounds.

Women catchers, dealers and commercial significance of tiger prawn

Prawn seed catching constitutes an important source of income for poor families in the Sundarbans. Surprisingly, prawn is also a source of quick and huge money for the islanders. For many Sundarbans islanders, prawn was like a lottery (Jalais 2010, 132) as it helped change their fortune almost overnight. In Bangladesh and West Bengal, the growth of an export-oriented brackish water prawn aquaculture industry started in the 1960s. The industry picked up very rapidly and by the late 1970s, prawn seed collection had become very popular in the Sundarbans (Ibid, 126). With the flourish of the industry, wind began to blow in favour of many poor landless families. The landless people found in this an opportunity to make cash. Even people involved in prawn collection and trading have bought land or house in the vicinity of Kolkata. Thus, women like Baishakhi who catch prawn seed in waist-deep water are not the only ones involved in this trade.

While travelling down the rivers in the Sundarbans one cannot help but notice women and children wading through chest or waist-deep water dragging nets to catch prawn seeds. They haul their nets along the riverbanks in the opposite direction of the current to catch the seeds that come with the tide. There are two main ways in which islanders catch or collect prawn seed. The first is when fishing families or groups set out on boats and cast their triangular nets in the middle of the river, pulling up one end of the net every one hour to transfer their catch into a container on the boat. The other method adopted mostly by women and children focuses on pulling nets along the bank of a river. A three or four feet wide and five or six feet long mosquito net fitted onto a wooden frame is what they use to collect prawn seed. However, this is a more predominant method and is more popular among women as it enables them to remain close to their neighbourhood and return to their household chores whenever they want to. Children on their way to school or way back find it convenient to spend time catching prawn seed along the riverbanks. Most of the poor families, who fish along the riverbanks, live in proximity to the river because their houses are located either on embankments or closer to the edge of embankments.

Once these women return to the villages with their catch, prawn seeds are separated and counted.[3] Prawn seed dealers who work for the khoti (little shacks built on embankments for prawn seed transaction) owners buy the seeds directly from these women and children. The dealers are often found strolling down the embankment waiting to buy the seed from the women fishing along the banks. It is in the presence of these agents of the khoti owners as buyers, that these women and children as sellers count and separate the prawn seeds they catch. The sellers with their trained eyes separate each individual tiger prawn seed of hair-thin diameter from the haul. The counting of prawn seed is a tiresome and tedious process with the sellers and buyers often picking up quarrel over the actual number of seeds counted and sold. While the dealers tend to bully the women catchers and pay them less, the women also retaliate, making the dealers pay the right price for the seeds.

The price of the prawn seeds is not fixed. The price varies depending on the season. During monsoon, prawn seeds are available in plenty. Women catch around 1,000 seeds per person every day. Because seeds are easily available, seed price comes down to Rs. 40/50 per 1,000 catch. However, during winter the seeds are difficult to come by. Women would be lucky if they could manage to catch 150–200 seeds per person per day. And the price soars up to as high as Rs. 500 per 1,000 catch. Sometimes availability of seed is greatly determined by the lunar cycle and wind direction. Experienced women equipped with the knowledge of right lunar cycle or nature of wind could catch more than what is expected during winter. Catching prawn seed along the riverbanks involves very hard labour. However, these women who largely come from poor households demonstrate enormous energy and endurance. After spending two to three hours in salt water, when their arms and feet turn pale and skins appear wrinkled, they seem ready to start their negotiations with the prawn collectors or dealers. Despite all their hard work, these women catchers seem to be at the receiving end of this transaction because the dealers or the collectors usually flex their muscle in fixing the price. The catchers are constrained to sell off their catch to ensure that the seeds do not die before they are sold. The catchers do not have the infrastructure to store these live seeds so as to sell them later when they think they are better able to negotiate the price.

Shrimp's significance and potential to earn revenue has been repeatedly emphasized by the fisheries department of the West Bengal government. In the departmental budgets submitted to the Legislative Assembly, the Left-front minister had stated that West Bengal was a major contributor to the total prawn production in the country (Nanda 1999b, 3). The annual budget proposals show how over the years the export value of prawn has been on the increase. It had increased from Rs. 55 crores in 1987–1988 to Rs. 354 crores in 1995–1996 (Nanda 1998, 3). It had further increased from Rs. 550 crores in 2004–2005 to Rs. 688 crores in 2009–2010 (Nanda 2005, 5; Nanda 2009, 5). Even after the change of left regime in West Bengal, the importance of shrimp has not lessened. The Trinamul Congress minister currently heading the department has stated that in the field of brackish water aquaculture, shrimp production has increased from 18,590 metric ton in 1990–1991 to 99,977 metric ton in 2010–2011 (Hena 2011, 5). It is clear from the successive budget proposals submitted by the department that out of a total export of fish, ninety per cent is shrimp (Nanda 2004, 4; Nanda 2005, 5). Thus, prawn is not only a source of livelihood in the Sundarbans, but it is also a lucrative business for islanders who act as dealers and fishery owners. Shrimp brings enormous foreign exchange for the government. Women like Baishakhi may not be aware of the export value of shrimp, but they are surely aware of the money that the shrimp market churns. They want to have their share no matter how insignificant that is when we consider the trade in its entirety. This explains their desperation to fish along the banks even when they know they run the risk of being killed by sharks or crocodiles.

Maya's voice and biodiversity, embankments and unsustainability of prawn catching

However, in an age marked by environmental consciousness and biodiversity conservation, prawn seed catching has become a matter of much concern among the international agencies, policy makers and non-state organizations for its perceived contribution to environmental degradation of the coastal common and mangrove vegetation in South Asia (Hein 2000; Sarkar and Bhattacharya 2003; Vidal 2003; IUCN 2009). In December 1996, the Supreme Court in India in a significant judgment ordered the closure of all shrimp farms set up within 500 meters of the high tide line and alongside creeks, backwaters, estuaries, rivers etc. (Sharma 1998). In the face of resistance from the prawn farming industries, an Aquaculture Bill was passed in 1997 which deviated from the Supreme Court decision and allowed the existing shrimp farms in the coastal zones to continue operations under a number of conditions (Hein 2000, 48).[4] While the fisheries department looks upon the prawn industry as being a source of revenue for the government, the forest department views catching of tiger prawn in great numbers as a threat to the ecosystem of the Sundarbans and as causing damage to marine resources and mangrove plantations (Directorate of Forests n.d., 4). The department has defined catching of tiger prawns along the riverbanks as a destructive form of livelihood (Directorate of Forests 2004, 15). Needless to say, women who pull nets along the shore are under the scanner. These women in pursuit of prawn seeds are believed to have been causing a great harm to the rich marine resources of the Sundarbans. When they pull nets along the banks in search of seeds, their nets catch many other marine lives in their larval form. But they tend to discard or throw away these marine lives in an attempt to collect the seeds. Many valuable marine lives which these women cannot identify as prawn seeds are killed in their larval stage.[5] Their frantic search for prawn seeds has put the valuable flora and fauna of the Sundarbans at stake. It is in the light of the increasing global concern about conserving biodiversity (Wilson 1995, 12) or portraying biodiversity as a global public good (World Bank 2003, 166) that women prawn catchers' livelihood appears as destructive of the ecosystem.

Prawn seed catching along the riverbanks appears problematic at another level. The women prawn seed catchers are believed to have posed a serious threat to the sustenance of the Sundarbans embankments. Women in search of prawn seeds keep going up and down the riverbanks, thereby disturbing the siltation process at the bottom of embankments. As has been mentioned earlier, embankments are fragile mud walls constantly eroded by river currents. At the same time, silts and mud carried by rivers tend to settle at the base of embankments. Thus, deposition of silt at the base of these embankments is crucially important for their sustenance. However, women prawn catchers' continuous movement along the shore is believed to unsettle the process of silt deposition, thereby weakening the embankment. Thus, these two perspectives – biodiversity loss and weakening of embankments – have not only gained currency in the corridors of governmental power and informed governmental policies, but also significantly shaped the orientation of NGOs working in the Sundarbans region. A discourse has emerged and gained ground that views prawn seed catching essentially in negative terms (Kanjilal 2000; Sarkar and Bhattacharya 2003; Vidal 2003; Badweep Barta 2007; Ajker Basundhara 2010). Voluntary agencies and NGOs working in the districts of South and North 24 Parganas subscribe to this discourse, embarking upon government-funded awareness campaigns making people aware of the importance of an alternative livelihood.

The issue of prawn catching reminds me of my encounter with Maya, a prawn catcher and a resident of Jagatpur, in a belligerent mood cursing her neighbour for the latter's allegedly adverse comments on prawn catching. It was early morning when I went to the riverside to take a few snaps of the embankment, river and the forest across the river. It was high tide and women were found drawing nets in knee-deep or waist-deep water in search of prawn seeds. While I was busy taking photographs, a man came and stood next to me. He introduced himself and began conversing with me. The man worked in a government office in Kolkata and was a resident of Jagatpur where he owned land and a house. He lived in Kolkata during weekdays and visited his home over the weekend. He was talking to me while we took a stroll down the embankment. He seemed displeased with the women prawn catchers. Being a government servant with a reasonably good salary, his views about prawn seed catching were shaped by the governmentalist discourse focusing upon biodiversity loss. He was particularly worried about the menace that prawn catching posed to the embankments and felt strongly that the problem of erosion could have been controlled substantially if fishing along the banks were stopped.

While he was expressing his views, Maya came from behind and asked him, 'What did you say?' Caught unawares the man was at a loss for words. Maya had a net in her hand and her soaked clothes suggested that she was among those who were catching seeds in the river. Pointing to me she told him, 'Just because this man is new to the village, you are telling him about the problems of prawn catching. Go and check if erosion has stopped in places where fishing has been stopped.' 'But you have not heard what I said', the man tried to say in his defence, but could not proceed further. Maya stopped him and asked, 'You earn a lot of money, but did I ask you to feed me? The next time you open your mouth, the consequences will be serious'. She was almost threatening in her demeanour. I knew Maya through Baishakhi. She was Baishakhi's neighbour. They both caught prawn seeds in the river. In fact, Maya was among those who rescued Baishakhi from the kamot. Maya had a small family with two children and husband, who worked as a construction labourer in Kolkata. She had to look after her children as her husband mostly remained away from home. The family migrated from Bangladesh. They had their house located on the path that ran parallel to the embankment. In the absence of any cultivable land, prawn catching and selling was a major source of survival for Maya and her family.

Beyond ecofeminism: Women prawn catchers and power relations within households

In view of the increasing involvement of women in the trade, a question that assumes significance is how do we construct women's role as prawn seed catchers? This question is important because a dominant perspective on South Asia – advanced by ecologists like Shiva and Mies – suggests that women should be viewed as contributing to nature or the environment (Shiva 1988; Mies and Shiva 1993). According to Shiva, a characteristic feature of Indian cosmology is the presence of feminine principles in nature (Shiva 1988, 38). Women are not only viewed as an intimate part of nature, but seen as related to nature in an essentially non-exploitative way. According to Shiva, this relation had been destroyed by the onslaught of western modernity. Western science and modernity, manifested in their institutional forms such as state and market, are viewed as having caused violence to nature in the name of development. The experience of women's participation in contemporary ecological movements in India has led Shiva to argue that because women embody nature, one can see in their protests a critique of destructive development processes and a possible return to India's pre-modern past where the problematic relations between sustenance and conservation would be resolved (Shiva 1986, 1988).

The depiction of women as closer to nature appears problematic, for it essentializes both women and nature. Such essentialized conceptions of women have been critiqued in anthropology (Ortner 1974; Gillison 1980; Ortner and Whitehead 1981). Gender and sexuality are viewed as cultural and symbolic constructs rather than natural facts (Ortner and Whithead 1981). Women, it is argued, are not any closer to or further from nature than men are (Ortner 1974, 87). Sinha et al. have questioned the perspective developed by Shiva and others such as Nandy (1988), Visvanathan (1988, 1991) and Alvares (1988) for essentializing Indian tradition (Sinha et al. 1997). It is argued that Shiva's and others' depiction of pre-colonial Indian society as marked by harmonious social relations and the absence of gender and environmental exploitation is unacceptable in the face of inadequate historical evidence (Ibid). However, Sinha et al. acknowledge that Shiva's approach has gained considerable tactical leverage in its struggle against modernity. Her portrayal of poor peasant women embracing trees to prevent their felling has become a global icon of popular protest against the degradation and exploitation of nature (Ibid, 66). A representation of this kind has also led multilateral agencies like the World Bank, which clamour for sustainable resource use, to portray women as natural conservers of resources because of their deep concern for the quality of the ecosystem (World Bank 1994, 28).

The question of whether women are closer to nature needs to be understood in the context of wider social processes of which women are part. In this respect, it is worth considering the views of Agarwal and Jackson who locate women's agency and their relation to nature in the context of domestic and agricultural divisions of labour (Agarwal 1988, 1989, 1994; Jackson 1993a, 1993b, 1998). However, in doing this, both move away from the ecofeminist perspective that sees women as the 'natural constituency' for conservation activities (Jackson 1993a, 1993b, 1993c) since it hinges on biological determinism (Jackson 1993b, 396; Agarwal 1992, 123; 1994, 37). According to Agarwal, the discourse that traces the connection between women and environment to female biology or symbolically identifies women with nature tends to neglect the material basis of the connections in the gender divisions of labour, property and power (Agarwal 1992, 126). To this end, she problematizes the view of household as a unit of congruent and unitary practices (Agarwal 1988, 83) and argues that the question of women's land rights has not been addressed in the literature dealing with women in poverty and development policies pursued by the Indian state. In this regard, the example of West Bengal is hardly encouraging, for under the Left-front government's land reforms programme tenants registered for land tenure were primarily men (Agarwal 1994, 9). In addition to the absence of land rights for women, there are intra-household gender inequalities in the division of labour and in access to food and health care. In the case of poorer households, the burden of poverty falls unequally on women, for linked to unequal economic status are inequalities in food intake resulting in low nutritional status, morbidity and mortality (Agarwal 1986, 1988, 1994).

Given the gender divisions of labour prevalent in the household, it is difficult to assume that women's practices reflect their choices or priorities. In this respect, Agarwal's views converge with those of Jackson, both of whom argue that women's concern for natural resource management is integrally connected to their livelihood strategies, especially the ways they are expected to function within the household. Thus, if women are adversely affected by large-scale deforestation and environmental degradation, it is because they depend on the forests for procuring the fuel and firewood necessary for cooking and feeding the family members, activities that are defined as 'naturally' suitable for women to perform. Jackson further argues that women's preference for dry and dead wood for fuel should not, as has been done by the ecofeminists like Shiva, be viewed as a reflection of their 'natural' concern for environment and forests, but rather dead or dry wood is lighter and easier to carry (Jackson 1993a, 1948). Those valorizing women's natural flair for nature conservation seem to ignore the fact that environmentally friendly management practices by women can often be explained in terms of pragmatic short-term interests, the roots of which can be traced back to women's position as defined within the household (Jackson 1993b; Agarwal 1992). Even on the agricultural front, there is a need to distinguish between the land farmed by women on their own account and that farmed as a part of household responsibilities when women are expected to supplement more as dependents than as autonomous individual actors. Under these circumstances, one needs to be cautious in asserting complementarities of women's gender interests and environmental interests (Jackson 1993c, 672) or interpreting women's engagement with land and forests as evidence of their enthusiasm for environmental conservation (Jackson 1993b, 407).

When we turn to the Sundarbans in the light of the analyses of Agarwal and Jackson, we find that women are engaged in similar activities for feeding and nurturing their household members. They too procure firewood from the forests and catch fish or crabs in narrow creeks to cook for family members.

Agarwal's argument about the absence of land rights for women also applies to the women of the Sundarbans. While working at the local land records office, rarely did I come across a land deed being registered in the name of a female member of the family. Very often women are victims of power relations obtaining in households. Married women are largely dependants of their husbands. Land belongs either to the husband or to the husband's family. Without land ownership rights, she largely remains powerless within the household. However, the condition of widows in the Sundarbans is even deplorable. Bhuli, a village in Gosaba block, had a locality called Bidhobapara (i.e., area inhabited entirely by the widows who lost their husbands to tigers). Bidhobapara was found located in proximity to the forest. What separated Bidhobapara from the forest was the river Matla that flowed southward towards the Bay of Bengal. This locality had about twenty-five to thirty widow households. The widows led a very vulnerable life as majority of them did not own any land. There were various narratives of land deprivation doing rounds in the locality. Many of the widows did not get their deceased husbands' share of agricultural land because their in-laws did not want to part with their family land. Some of them continued living with their in-laws but they were asked to earn to raise their children. Others who could not continue in their in-laws' house because of humiliation and ill-treatment, managed to build a separate house adjacent to the in-laws' house. There were still others who were denied access to the agricultural land that was gifted by their parents as dowry during their marriage. The widows were denied their claim to the land on the ground that the land given as dowry was gifted to the families of sons-in-law. The widows were given to understand that they could not stake their claim to a property gifted by their parents at the time of marriage. Many widows worked as wage labourers to raise their children. However, for majority of these women, river remained an important source of livelihood. To earn their living or even to supplement their meagre income from working as wage labourers, most of these women caught tiger prawn seeds in the river.

In many ways, Agarwal's and Jackson's accounts transcend the simplistic equation established between women and nature and help us understand the complexity of women's position and agency within the household. The argument that women's relation to nature needs to be grounded in material processes and immediate livelihood strategies seems particularly relevant when we turn to women's engagement with prawn catching in the Sundarbans. Prawn catching along the riverbanks is predominantly a woman's occupation in the

Sundarbans, for three to four hours of prawn catching is believed to fit in well with the other domestic chores of women. Women catch prawns in various capacities, as 'natural' dependents, when their role is one of supplementing family income, as heads of households where the husband is dead or as temporary heads when husbands are away from home. However, in each case, the overriding concern seems to be one of caring, nurturing and feeding the family, a role whose roots seem to lie in the way woman's position is defined within the household. It is not simply the absence of land rights, but also factors such as domestic power relations, early widowhood, levels of support from offspring and kin, which account for women's hardship (Jackson 1998, 46) and also mediate their relation to nature.

Thus, the above discussion focusing on the plight of the widows serves to show how land relations both within and outside the household dominate, marginalize and deprive women of their legitimate share. However, this chapter, as has been mentioned earlier, is not concerned with land, land relations and agriculture. Rather, this chapter focuses on the centrality of shrimp farming and shrimp fisheries or bheri[6] in the Sundarbans. In the centrality of prawn farming and fisheries lies the peripheral significance of agriculture. In a region struck by land erosion, saline ingress and disastrous cyclones, agriculture remains of marginal significance. This chapter shows how ecology and shrimp fisheries are constitutive of each other. The chapter shows how the embankment is central to the understanding of prawn farming in the Sundarbans and how the embankment remains a pivot around which revolves shrimp trade and shrimp politics.

The significance of prawn catching also lies in the way it has provided women with the chance to move out of the impasse created by the absence of their rights to land on the one hand and the peripheral position agriculture occupies in the life of the islanders on the other. Unlike agriculture, prawn seed catching is not seasonal, for with every high tide come thousands of prawn seeds. Women catching prawn seeds do not have to think about the market because immediately after the seeds are collected, the dealers and traders wait to buy the catch and take them to the fisheries where they are reared. Women thus emerge as constrained but competent social actors capable of articulating their priorities (Kabeer 1994). However, women's participation in the trade is not unique to the Indian Sundarbans alone. Ito has also noted the increasing participation of women in activities around prawn farming in Bangladesh (Ito 2002, 63–65). In writing about the people living on char – low-lying sandy masses that exist within the Damodar riverbeds of lower Bengal – Lahiri-Dutt and Samanta observe that livelihoods tend to become gendered (Lahiri-Dutt and Samanta 2014, 189). Unlike women from the Bihari community who never go to fish in the river, Bangladeshi women from families living on char fish in the river for subsistence (Ibid). Apart from meeting subsistence needs, women sell their catch as well. Bangladeshi male folk consider it beneath their dignity to sell fish to households and delegate this job to their wives (Ibid).

Prawn trade and centrality of bheri

However, women are not the only ones engaged in prawn trade in the Sundarbans. As mentioned earlier in this chapter, there exist networks of catchers, collectors, dealers and traders involved in the trade. Similar networks are found to exist around fish or prawn seed cultivation in Bangladesh as well (Lewis et al. 1993; Ito 2002). I have already described how the dealers or khoti owners buy tiger prawn seeds from the women catchers. Each khoti has a register in which each dealer or agent of the khoti owner notes the number of seeds bought, the name of the catcher from whom the seeds are bought and time and date of the transaction. Once the seeds are collected, they are kept in the khoti and stored in containers filled with water until they are sold to the owners of bheri or fisheries. Khoti owners are not big businessmen; they are small dealers who often set up their khoti in partnership. The dealers, who buy seeds from women, could be partners who have jointly set up their prawn dealership by pooling their own individual money. They do not have the adequate infrastructure or provision either to store, rear or groom these seeds before they mature into tiger prawns. They serve mostly as middlemen standing between these women and bheri owners. After the seeds are collected by the dealers, they are poured into containers placed inside the khoti. Fishery or bheri owners come to the khoti and buy these seeds. They do not count seeds from each container. Rather, they pick one of the containers at random and count seeds from that container. Prawn seeds travel to bheris where they are reared and groomed into tiger prawns for their export and sale in the market. Thus, bheris are the pivot around which prawn catching, farming and trading revolve in the Sundarbans. These bheris are the nerve centres of the Sundarbans, the very constituents of the landscape and topography of the delta. While travelling on a bus through places like Bhangar, Ghatakpukur, Malancha, Sarberia and Sonakhali one cannot miss these fisheries on both sides of the road. These are square or rectangular shaped water bodies separated from each other by earthen mounds built around them. Some of these bheris are so big that their boundaries become invisible. Standing in front of these bheris, one would experience a huge expanse of water whose boundary seems to disappear into the horizon. In these fisheries, seeds are kept with considerable care to grow into prawns, one of the biggest sources of foreign exchange for India.

Some of the biggest fisheries are found located in Ghatakpukur and Malancha. Here, we come across fisheries with land area measuring up to 300/400 bighas. These are plots of agricultural land converted into fisheries. However, bheris are not only confined to Ghatakpukur and Malancha alone. Fisheries are spread across the length and breadth of the Sundarbans, because every now and then agricultural lands are being converted into bheris. This is where we stand face to face with ecological peculiarities of the region and the contingent nature of people's livelihood practices. With the beginning of tiger prawn seed catching and shrimp trade in the 1970s, bheris have begun to proliferate. Today, the Sundarbans blocks such as Basanti, Gosaba, Patharpratima, Sandeshkhali, Haroa have fisheries dedicated to shrimp farming and trade. These bheris are not certainly as big as those found in Ghatakpukur, Malancha or Sarberia, but it is not difficult to come across relatively large fisheries occupying an area of 100/150 bighas. However, it is not the popularity of prawn seed that alone accounts for the presence and proliferation of bheris in the Sundarbans. As has already been mentioned, the existence and continuance of bheris has to do with the uncertainty and peripherality of agriculture.

When disaster strikes the Sundarbans, agricultural land and crop are the first to be badly affected. Crops are completely destroyed as happened in the cyclone of 1988. The cyclone happened in the month of November when the rice fields had fully grown crops. For all practical purposes, land became unsuitable for cultivation in the subsequent years. The cyclone spelt untold misery for farmers, for years and months' of effort went down the drain in a few hours. History seemed to have repeated itself when cyclone Aila struck the Sundarbans many years later. I already discussed the devastation the cyclone caused and the impact it had on land and agriculture. People began to leave the Sundarbans because one major source of livelihood, that is, agriculture became unsustainable. Agriculture proves unsustainable even during the period between cyclones. As stated earlier, the continuous erosion of land caused by rivers flowing in a particular manner is what makes the embankment vulnerable. As a result, water seepage occurs through the cracks and breaches that develop in the embankment. Saline water salinates rice fields and ponds. The situation worsens when a stretch of a vulnerable embankment collapses resulting in immediate flooding of lands, fields and houses. Salinity destroys the prospect of sustained agriculture. Thus, agriculture remains a peripheral activity in a region struck either by cyclone or by a more imperceptible process of land erosion and embankment collapse. In desperation, farmers convert their lands into fisheries.

However, farmers are not the ones who start fisheries. Rather, they lease out their lands to one or more than one individuals (a lessee or lessees) who buy farmers' lands roughly for a period of five years.[7] Generally, it is a collective decision by a group of farmers to lease their lands to individuals who are prospective owners of the proposed fishery. No single farmer, unless he has a huge land at his disposal,[8] could lease out his land for the construction of a bheri, because a small plot of land ceases to be an effective fishery from a commercial point of view. The fishery owner pays lease money at the rate of Rs. 3,500–4,000 per bigha of land to be brought under pisciculture. Thus, it is always a case of about 10–12 farmers (sometimes even more) coming together and leasing their individual lands to the fishery owner. Their individual plots, small or big, put together, would become a reasonably big plot for building the fishery. A deed of agreement is signed between the owners of the lands (farmers) and the owner of the fishery (the lessee), where each of the farmers' contribution in the form of his plot of land, its measurement, how much of money the fishery owner would pay to each farmer and in what instalments are all mentioned. Once the agreement is signed, the fishery owner is responsible for all payments to the farmers and the building up of the fishery.

Location of the fishery land is equally significant. If a fishery land is found adjacent to the river, building that fishery becomes much more convenient. The fishery needs to be connected to the river. If a fishery is located away from the river, a canal needs to be excavated connecting the fishery to the river. This connection is important because saline water needs to come in and go out of the fishery during high and low tides, respectively. Once the connection is established, the fishery owner builds earthen walls around the fishery land to demarcate its boundary. A sluice gate is placed at one end of the fishery wall to control the flow of river water into the fishery. Once the walls are erected around the fishery, the land is emptied of water and treated with lime and medicine to ensure that prawn seeds remain free of diseases. After the ground is prepared, the sluice gate is unlocked to release the river water into the bheri. Fishery owners buy prawn seeds from the dealers who directly collect them from women who catch the seeds in the river. The seeds are released into the bheri at the rate of 1,000 tiger prawn seeds per bigha of fishery land one owns. For example, if fifty bighas of land is converted into a fishery, roughly

50,000 prawn seeds are likely to be put in the fishery. Initially, a small reservoir called nursery is created within the fishery where the seeds are kept until they grow to a reasonable size. The reservoir is separated from the main fishery by earthen walls built around it. When the seeds reared in this reservoir become sufficiently big, a wall of the reservoir is broken to allow the tiger prawns to swim into the main fishery. It takes about two to three months for the seeds to become tiger prawns. When the prawns are fully grown, they are collected in a fish trap called patan, placed in front of the sluice gate of the fishery. The fish trap looks like a basket. Patans are mainly of two types: one is made of bamboos and the other is made of mosquito net mounted on wooden frames. During high tide when tidal water enters the bheri through the sluice gate, the tiger prawns are drawn towards the sluice gate against the tidal current. They enter the patan and get trapped inside. When a sufficient number of prawns are collected in the patan, it is emptied and the prawns are transferred to a container. Thus, prawns are ready for sale in the market.

One is likely to come across a hut built near the fishery. It is called alaghar where a register to record the details of the trade and fishing repertoire such as net, fish trap and fishing containers are kept. This hut is called alaghar because it is built on the aal or the earthen mound that serves as the boundary of the bheri. Sometimes the hut is built inside the bheri on bamboo stilts above the water. A fishery owner needs to appoint security staff to keep a vigil on the bheri. Each security staff is paid a monthly salary of Rs. 4,000 with two to four holidays per month. The security staff needs to keep a vigil on the fishery at all times. This is important in view of criminal activities ranging from petty theft to poisoning of the fishery. A fishery owner need not be someone from the villages of the Sundarbans. Often the owner is an outsider who comes from Kolkata to buy a fishery on lease. However, what is significant to note is that he has to employ local people in the running of the fishery. Thus, the owner buys prawn seeds from the local dealers, employs villagers for the day–to-day management of the fishery and also sells tiger prawns to the local sellers or paikars who in turn sell prawns to Kolkata-based businessmen. Local people are involved in various stages of prawn catching, farming and trading in the Sundarbans. While for some it is a source of money and profit, for others it is livelihood and employment.

Prawn trade, prawn politics and local newspapers

What I have described above are the mundane details of the working of fisheries and who stands where vis-a-vis prawn catching and trade. However, what I narrate in the following pages is the unfolding of politics surrounding fisheries and prawn trade which is often found transgressive of the so-called formal rules of the trade. Prawn fishery and trade often figure in significant ways in the local newspapers published from the Sundarbans. I have already introduced this local fortnightly newspaper *Badweep Barta* in connection with our discussion of the cyclone Aila aid and relief in Chapter 3. I present here a few news reports, focusing on the trade. Highlighting the adverse impact that prawn farming has on the Sundarbans embankments, *Badweep Barta* reports,

> There have emerged in recent times many illegal fisheries or bheri throughout the Sundarbans. For the purpose of pisciculture, river water is released into these fisheries. This results in the weakening of the Sundarbans embankments (Nayek 2005; translated from vernacular).

Badweep Barta in its newspaper dated 16–28 February, 2007 carried another interesting report which states,

> There has grown in the past one decade or more a tiger prawn–dependent economy in the Sundarbans. It is a source of quick cash. This has suddenly transformed the economic condition and consumption pattern of many families in the Sundarbans. People's greed has led them to become extravagant and wasteful. Tiger prawn is a marine resource of the Sundarbans. This resource needs to be used or exploited in a planned manner so as to ensure that it remains a stable source of livelihood for many families. Such planned resource use will help counter a major problem posed by indiscriminate building of fisheries. Agricultural lands are being taken on lease to transform them into fisheries. There are deliberate attempts to release saline water into agricultural lands adjacent to a fishery, making lands unsuitable for cultivation, thereby converting them into a fishery (*Badweep Barta* 2007; translated from vernacular).

If the above reports have stated only in generic terms the harmful implications of prawn trade and hinted at the presence of vested interests in the building of bheris, the following report in *Badweep Barta* shows how prawn trade becomes a locus of fierce and vengeful political rivalries. The report states,

> Recently, CPI-M and Trinamul Congress activists have locked horns over an illegal bheri in Chunakhali of Basanti block. As a result, five houses in the locality have been burnt. At a protest meeting organised on 25th February by

the CPI-M, Iudali Sheikh observed that local Trinamul Congress cadres have threatened to kill Wahed Ali, the head of village panchayat of Phulmalancha. The CPI-M leader alleged that it was under the guidance of local Trinamul Congress leadership that saline water was deliberately released into agricultural land in Baguakhali village to turn about 1,000 bigha of land into an illegal bheri. At the meeting, the local CPI-M leadership insisted on organizing a movement to protest against Trinamul Congress cadres' tyranny, panchayat corruption and saline water fishery. The local RSP leadership also reiterated the need to ban illegal bheri (*Badweep Barta* 2003; translated from vernacular).

Thus, prawn fisheries are a bone of contention. They are a source of fierce battles among political parties with each trying to extend its base into other's territory. They are a locus of patron clientelism. Political leaderships help set up fisheries, thereby ensuring the loyalty and allegiance of prawn dealers and fishery owners. The burning of houses or life threats mentioned in the above report are consequences of such rivalries and conflicts. While the above report mentions life threats being issued to the CPI-M panchayat head of Phulmalancha, another edition of *Badweep Barta* reports the killing of a prawn seed dealer who was a Trinamul Congress cadre. The report states,

On the night of 28th December, miscreants killed a Trinamul Congress worker. … The name of the deceased was Swapan Mondal. From family and police sources it was known that on the night of 28th December, Swapan left home to buy tiger prawn seeds and never came back. Next morning his body was found adjacent to his house with his throat slit (*Badweep Barta* 2012; translated from vernacular).

The reports above have hinted at the growth of illegal fisheries. A question that arises is why the reports dubbed bheris as illegal, especially when we know that leasing of fisheries is done through an agreement between the farmers and the lessee. The answer is that it is by breaking the embankment that fisheries are made. Saline water is released into the agricultural lands to turn them into fisheries. Unless a stretch of an embankment is deliberately broken, water cannot be released. This is often done in defiance of the irrigation department, which owns the vast stretch of embankments in the Sundarbans. Rarely, if ever, is permission sought from the irrigation department before a fishery is planned. Building fisheries is as much a part of the pre-Aila Sundarbans as it is of the post-Aila, the only difference being that since cyclone Aila tiger prawn fisheries are significantly on the increase. Fisheries are integral to the land and livelihood of the Sundarbans, so much so that in the making and unmaking of fisheries the thin line of distinction between legality and illegality appears fuzzy.

When a stretch of embankment collapses, the land lying in proximity to the embankment is the first to get flooded. If the embankment collapses during high tide, a considerable area gets flooded instantly. It is only during low tide when the water retreats that the villagers first try and repair the broken embankment. If the repair work proves difficult in the absence of a work order from the irrigation department, then, judging the progress and direction of saltwater ingress, the villagers build temporary protective mud walls inside the villages to prevent water from making further inroads into areas which were not yet flooded. Subsequently, the irrigation department acquires land to build a new embankment behind the old stretch that has collapsed. This implies further acquisition of village agricultural land to build the new stretch of embankment. The acquired land, which was earlier part of the village land now falls on the riverside, and becomes government's property. With no compensation from the government, victims of land acquisition are given an opportunity to convert the acquired land into a source of livelihood. However, victims remain aggrieved for their lost lands and this sense of deprivation often leads them to break embankments and deliberately release saltwater into neighbouring lands, thereby destroying lands which still remained part of the village agricultural land and had not yet been acquired by the irrigation department.

Niranjan as prawn dealer and politics of bheri making

Niranjan was a prawn seed dealer who lived in Jagatpur village on Kusumpur island and he had his khoti on the embankment path that ran parallel to the riverbank where women drew nets to catch seeds. This was the place where I met Maya, the prawn seed catcher in her belligerent mood. I met Niranjan in his khoti on a number of occasions and we had long conversations. It was during one such meeting that Niranjan stated that tiger prawn was the very backbone of the Sundarbans and bheris were instrumental in the making and unmaking of embankments. When I asked why he thought so, Niranjan said that breaking of embankments and making of fisheries were complementary to each other. When embankments collapsed and the irrigation department acquired land to build new embankments, people who lost their lands became revengeful, because their own lands were lost to the river. Even when lands were converted into a fishery, individual land became part of a collective space. It was this sense of loss that led them to break embankments and destroy others' lands as well. It was as if fishery was the inevitable destiny of all lands in the Sundarbans. Illustrating the case of Garantala embankment collapse where the irrigation department acquired thirty-six acres of land (discussed earlier in Chapters 3 and 4), Niranjan asked me, 'Do you know why the embankment in Garantala collapsed?' 'The embankment became weak and it collapsed during high tide', I replied. 'What you heard was only the tip of an iceberg', Niranjan said sarcastically. Niranjan went on to state the following:

> There existed a small fishery close to where the embankment collapsed in Garantala. This fishery was the result of a previous embankment collapse in that area. Only three or four farmers who lost their lands on account of land acquisition set up and ran the fishery on their own. They could not lease their land to help build a fishery because the land lost was not big enough to attract the attention of someone who could make profit out of it. For want of money, farmers found it difficult to run the fishery. Moreover, because of continuous seepage of fishery water, agricultural lands bordering the fishery were found contaminated and salinated. Owners of the neighbouring lands picked up quarrel with these fishery owners almost every day. Once while working in their fishery they noticed huge cracks in the embankment, a few meters away from their fishery. They were so revengeful and fatalist that they did not bother to raise an alarm, but simply waited for the cracks to become wider. By evening when the river rose during high tide, the cracks became huge holes with water flowing at a great velocity. In no time the tidal wave broke open the embankment, resulting in the flooding of a wider area. The villagers did this only to ensure that others would lose their lands exactly the way they did.

'So it was a complex issue, a combination of factors', I commented. 'But then what did irrigation do when all this happened?' I asked further. 'Irrigation eventually acquired land and built a new embankment. Many lost lands. Engineers and contractors made money and gave money to some of the panchayat leaders to ensure they remained part of the engineer–contractor network. The villagers who lost lands were given an opportunity to set up a fishery. A moneyed man from Kolkata has set up the fishery. He runs it successfully because he has money. He pays visits to the irrigation office in Gosaba to ensure that engineers do not come over to inspect his fishery'. Niranjan did not even pause for a breath. 'But how could those villagers be so callous and irresponsible? Don't you think it was suicidal on their part not to raise an alarm and prevent the disaster?' I was curious. 'There's no point blaming them. They lost lands and the fishery they were running was unprofitable. Yes, it was suicidal not to raise an alarm, but they wanted to ensure that others suffered the way they did'. Niranjan sounded a bit self-righteous. In Chapter 4, I discussed how the agency of embankment workers was manifested in various acts of subversion and resistance. Here, we also encounter subversive acts centering around making of bheris. There are various rationalities at work explaining people's desperate search for livelihoods in a region where livelihood opportunities are difficult to come by. Salt water and prawn fishery become a great leveller, ironing out the differences between those who still own land and those who stand deprived of it.

Here, once again we revisit the riparian rules I discussed in Chapter 4. When land is acquired for re-building embankments, villagers lose their ownership over the lands lost due to the building of a new embankment. Individual villager's right to the lost land bordering the river (where the new stretch of the embankment is erected) is inconsequential considering the salinity of rivers in the Sundarbans. In the absence of individual right and compensation for the lost land, villagers losing land are given an option to collectively use the land bordering the river as a fishery. However, a small plot of land does not make fishery a commercially viable enterprise. The above incident – where the villagers on discovering the embankment breaches paid scant attention to them, pinning their hopes on the river to do the rest to inundate larger agricultural fields so as to ensure that others' lands could be converted into fishery – shows how collective rights to fishery could only be established through recourse to violating individual rights to agricultural land.

Cross-border prawn trade and networks

Niranjan was in the prawn trade for quite some time now. He had set up his khoti in partnership with Subol who lived on Garantala island right across the river. Every morning Subol came by boat to Jagatpur. He left his home on his bicycle to reach Garantala ferry where he got his bicycle on the boat and crossed the river in ten minutes to reach Jagatpur ferry. He got his bicycle off the boat and drove it to reach the khoti. They sold seeds to local bheri owners who came from different villages of Gosaba block. They also sold seeds to dealers who took them on boats to Satjelia or Basanti market where bheri owners congregated early morning to buy prawn seeds. Sometimes Niranjan or

Subol even travelled to Malancha or Ghatakpukur to sell these seeds to large bheri owners slightly at a higher rate. Because I spent considerable time in their khoti, I became familiar with people who visited the khoti to buy prawn seeds.

Early morning, once when I went to their khoti, I stumbled upon a few unfamiliar faces. The khoti was pretty crowded with Niranjan and Subol busy counting prawn seeds for their buyers. These new buyers were a bit aggressive in their tone. It looked as though they were bullying Niranjan and Subol. Once the counting was over, seeds were transferred into big aluminium vessels. A van rickshaw carrying the containers left the khoti. The transaction lasted for about half an hour and the crowd gradually dispersed. 'Are they your new customers?' I asked Niranjan and Subol. 'Yes, they are TMC people who got into the trade. We are under pressure to sell prawn seeds to them', Niranjan answered. 'Yes, their gestures suggested that they are influential people, but then who is compelling you to sell seeds to these people?' I further asked. 'TMC is now very powerful in Gosaba, the party having dislodged the RSP and the CPI-M is now trying to control prawn trade as well', they replied. 'Do these people own bheris or are they just dealers?' I was curious. 'They are dealers, but they are going to sell the seeds in Bangladesh, not here in the Sundarbans'. 'Bangladesh, how?' I exclaimed. 'Can you put me in touch with any one of them? I want to know how seeds are sold across the border', I insisted. Niranjan came closer to me and said, 'The people you saw are not the only ones selling prawn seeds in Bangladesh. These dealers form a network that operates throughout the Sundarbans. They are very secretive about their trade. They will not divulge it to anybody, except their close confidants. We know how these seeds are taken to the Bangladesh border'. Both Subol and Niranjan narrated how prawn seeds travelled to Bangladesh.

These dealers set out in small groups. They keep changing their travel plans. Sometimes they travel by a bus to Dhamakhali where they transfer their containers onto a boat and travel to Hasnabad, which is closer to the Bangladesh border. This is not the only route they follow to reach Bangladesh. At times they undertake a boat journey from Sonakhali in Basanti block and cross Raimangal river in North 24 Parganas to reach Basirhat, another port of entry to Bangladesh. Before reaching Hasnabad or Basirhat, they quickly replace big vessels carrying prawn seeds with small containers to avoid the boat being an object of attention. On arrival at the port of entry, Hasnabad or Basirhat, they inform their counterparts in Bangladesh. The Bangladeshi traders often operate through brokers or middlemen who wait at Hasnabad or Basirhat to meet the Indian dealers and help them cross the border by bribing the Indian

Border Security Force [BSF]. When the Bangladeshi brokers are not available, the Indian prawn dealers themselves cross the border by bribing the BSF. Once they cross the border, they sell their seeds to Bangladeshi traders. The transaction is mostly done in Indian National Rupee [INR]. But even when business is done in Bangladesh currency, money changers [persons carrying both currencies] are instantly available to convert them to Indian currency.

'This is a well-orchestrated expedition, I have to say. They ensure that everything works out their way. But then why do they sell seeds in Bangladesh?' I could not contain my amazement. 'The seeds are sold at a much higher rate in Bangladesh. I don't know the exact rate, but it's almost double the rate at which it is sold here. Moreover, those who sell in Bangladesh get more money, because the exchange value of Indian rupee is higher than that of Bangladesh taka', Niranjan replied.

About a week later, Niranjan took me to Basanti market in Sonakhali. Niranjan wanted to give me a tour of the early morning prawn market in Basanti. We crossed the river by a boat and then went on a van rickshaw to reach Basanti market. We set out very early and by the time we reached it was six o'clock. The market was crowded with tiger prawn seeds being sold in one corner and fully grown tiger prawns in the other. Big containers carrying five to six inches long bagda were being brought in to be auctioned to paikars or buyers who would sell them to Kolkata-based businessmen. They would be weighed on the weighing scale and given to the highest bidder. Because it was monsoon, bagda prices were on the lower side. Still bagda was being sold for Rs. 600–700 per kilogram. There were other fish available for sale, but bagda certainly stole the show. As I watched the business transaction intently, Niranjan drew my attention to a group of people smoking and chatting among themselves in one corner. Pointing towards them, Niranjan said, 'They are another group preparing to go to Bangladesh'. 'How do you know?' I asked. 'I know them', he said. Then pointing to a set of small aluminium pots stacked together, Niranjan said, 'They will transfer their prawn seeds to those containers when they reach Bangladesh border'. 'This is a well-established network. Why are you then wasting your energies here in Gosaba, why don't you also sell your seeds in Bangladesh?' I suggested. Thinking that I was being sarcastic, Niranjan said, 'You must be joking, one has to be part of the networks that organize and control the trade. And you need to have the weight of the party behind you'. 'By party you mean the TMC?' I asked. 'Yes, now it's the TMC. If we travel to Bangladesh and stop selling seeds to the TMC people here in our khoti, our existence will be in peril', Niranjan replied.

Tiger prawn sale is not the only trade one encounters on the India–Bangladesh border. Every day people ferry various goods including cows and goats across the border. Apart from economic activities, cross-border migration from Bangladesh to India is a recurrent phenomenon. The instance of prawn dealers bribing the BSF personnel suggests that borders as boundaries of states are internalized by these traders. Yet, borders remain, to use Ghosh's (2001) phrase, the shadowy lines, lines that exist only on paper. Borders are just not lines in the landscape, they denote a spatial dimension of social relationships that are continually configured; in this process, the meaning of borders is produced, reconstructed, strengthened and weakened (Basu Roy Chaudhury 2011, 40). Samaddar (1999) in his work argues that borders are fluid and contingent in nature and people often cross artificial political borders in pursuit of their social, familial or economic interests. And in doing this, they challenge the idea of finality of political borders. Border remains an uncertain zone, where the state confronts rationalities other than those of the state. Border is uncertain because it situates itself on the uncertain demarcation of the external and internal and continuously disrupts the authorizing presence of a power (Samaddar 1999, 20). Banerjee argues that border making is an artificial exercise. It was colonial rule which converted frontiers into borderlands and the partition was instrumental in transforming them into borders (Banerjee 2010, xxxiv). Borders, border-making and securitizing border are chief features of state formation in South Asia.

Prawn politics and Niranjan's predicament

Niranjan had reasons to be anxious. He was an RSP party worker, and with the help of the party, he set up his khoti business. However, with the RSP becoming a weak force in Gosaba, Niranjan found it difficult to get on with his business. He got into a partnership with Subol for strategic reasons. Subol was in close contact with the TMC party and panchayat leaders in Garantala. There were factional rivalries within the TMC. Niranjan thought Subol's contact with one such faction in Garantala would come in handy in their dealing with another faction in Kusumpur. Niranjan and Subol were not the only ones whose fortunes were made and unmade by the tidal waves of political change in the Sundarbans. In the last panchayat elections in 2013, the TMC's electoral strength in Basanti and Gosaba blocks had sharply increased. Out of thirteen village panchayats in Basanti block, the TMC captured ten. In Gosaba block, the TMC was in control of twelve out of fourteen village panchayats. In Basanti panchayat samity, the TMC had formed the board winning twenty-six out of thirty-nine seats, while in Gosaba panchayat samity, the TMC captured twenty-nine out of forty-two seats. The remainder of seats, needless to say, were shared between the RSP and the CPI-M. When we compare this electoral result with the 2009 panchayat election results in Basanti and Gosaba blocks (this is presented in Chapter 3 in connection with my discussion of the disaster-related aid and relief in the post-Aila Sundarbans), we realize the extent to which the TMC had made inroads in these blocks. However, the TMC's political penetration is not confined only to Gosaba and Basanti alone. These two blocks are only samples indicative of the TMC's growing political presence in the other Sundarbans blocks as well.[9]

This changed political scenario has contributed to intensifying political conflicts and rivalries in the region. Prawn trade and bheri lie at the centre of these inter-party or intra-party rivalries and clashes. The post-Aila Sundarbans has witnessed a considerable rise in the number of bheris. They are the biggest source of election funding for the political parties such as the TMC, RSP or CPI-M. Parties make bheri owners pay huge sums of money to meet election expenses. Fishery owners give money to enjoy political protection and patronage. Fishery owners who hail from the Sundarbans belong to one political party or the other or tend to align with the dominant party. Apart from giving money to political parties, bheri owners are also required to meet expenditure involved in carrying out election campaigns which often range from funding parties' public meetings to entertaining polling officials while they are on election duty. In this context, I present an interesting excerpt from the RSP's election campaign in Basanti block. As an RSP party worker, Niranjan was involved in the election campaign and had stayed in Basanti until the election was over. As an election campaigner, he witnessed many significant events. Among those, he chose to share an incident he thought worth remembering. He narrated the event as follows:

> The RSP deployed its party machinery [cadres, funders, fund raisers etc.] to ensure that its party candidate won the election. Among the funders, there was a Kolkata-based bheri owner who had his fishery at Jharkhali in Basanti. He was asked to take care of the polling officials who would be on their official duty to conduct election at a polling centre in Basanti. The fishery owner was made responsible for providing them with hospitality services during their stay. The rationale was to ensure that the officials remained so blissfully happy that they turned a blind eye to the RSP's electoral malpractices. The fishery owner lived

up to the party's expectation. His vehicle brought these officials from Canning to Sonakhali in Basanti. He escorted them to his fishery house in Jharkhali. On their arrival they were treated to tiger prawn fry and whisky followed by a dinner comprising of prawn fry, prawn curry made with coconut milk and chicken. Next morning they were given breakfast consisting of prawn cutlets, bread and eggs. In the afternoon lunch arrived at the polling centre. They ate fish, meat and tiger prawn curry. In the evening the fishery owner arranged for a launch to give the officials a tour around the Sundarbans forests. Whisky and prawn fry were served on board and finally, their eventful day ended with a sumptuous dinner. The fishery owner played his role brilliantly and the party realized its objective. The officials conducted election following the party's diktat.

'This is extraordinary. You seem to remember every bit of what they did and ate', I commented once Niranjan finished narrating. 'Why not, they ate while we watched. How many of us can afford to eat or even taste bagda? None of us, because it is so high priced. However, it is us who are engaged in bagda trade throughout the year. Embankments are broken, bheris made and lives lost in search of bagda. The irony is that we risk our lives and catch bagda, but bagda remains elusive for us. Many families haven't had rice ever since Aila happened, because rice has not grown in the field', Niranjan was angry. His spontaneous outburst was what made his story so very poignant. 'But now it's the TMC, the RSP or CPI-M regime is almost over'. I deliberately wanted to change the course of our discussion trying to draw Niranjan's attention to something more mundane. 'Yes, now it's the TMC and people are joining the TMC in large numbers. The fishery owner who worked for the RSP has now joined the TMC', Niranjan replied. 'Really', I exclaimed. 'Not only bheri owners, even bagda dealers and traders have started to move towards the TMC. But people are joining and getting killed as well'. 'Why are people getting killed?' I asked even before Niranjan completed. 'Just before the last parliament election, a bagda dealer in Gosaba got killed. I knew him personally. He was a CPI-M cadre and close confidant of a CPI-M leader in Gosaba. He was an industrious and energetic person. He left the CPI-M to join the TMC. Within a few months of his joining the TMC, he got killed'. Niranjan narrated the incident:

Pintu[10] was a prawn seed collector and a CPI-M worker. Before the last parliamentary election he joined the TMC. A faction of the TMC had been putting pressure on him to join the party. This faction, largely controlled by the prawn seed dealers, was in conflict with the other faction of the TMC. Pintu joined the TMC and was very active in raising funds for the party. He

got a lot of money from the prawn dealers' networks. This was not to the
liking of the opposing faction. The members of the opposing faction thought
that the prawn dealers' faction had employed Pintu to discredit them. So they
killed him. At night when Pintu was returning home a group of people came
and stood in his way. He was first beaten and then his throat was slit with
his hands tied to a tree. TMC blamed his death on the CPI-M. But it was
actually a rivalry within the TMC that claimed Pintu's life.

At the end of his long narrative, Niranjan paused for a breath. He
looked as though he was lost in his own world. He was angry, yet his anger
had a trace of sadness in it. His sadness resulted largely from a sense of
uncertainty surrounding life and livelihood in the Sundarbans. Niranjan was
at his metaphorical best when he said, 'We risk our lives and catch bagda,
but bagda remains elusive for us'. Nothing could capture this uncertainty
more succinctly than the above words of Niranjan. Despite lifelong search
for a livelihood, livelihood remains elusive for the majority of people like
Niranjan. We stand face to face with lives that remain perpetually unsettled
in the delta.

Political society, prawn and the state: Beyond sustainability-unsustainability

Following tiger prawn, one enters upon a society in which people pursue
their livelihoods amid conflicts, rivalries and pursuit of vested interests.
Chatterjee (2004) employs the term political society to comprehend the nature
of popular politics in recent times. The notion of political society problematizes
our conventional understanding of state and society as dualistic categories.
Thus, politics is no longer played out in a state–civil society mode where
right-bearing citizens inhabiting civil society influence state policies. Rather,
the domain of popular politics is shaped by increasing negotiations between
population and governmental technologies that the state deploys to provide for
the welfare of the population. Governmental technologies bring population
into a certain relationship with the state. This relation, according to Chatterjee,
does not conform strictly to principles enshrined in the constitution. In other
words, the relation could be found transgressive of the formal boundaries of
constitutional politics. I discussed the idea of political society earlier in Chapter
5 where I examined concepts such as indigeneity and community in the light
of adivasi beldars and their practices. I further draw on the idea of political
society here in this chapter to deepen our understanding of prawn politics that
unfolds in the Sundarbans.

What we encounter here is not a community of fishermen, but a collective of prawn catchers, collectors or dealers, bheri owners and paikars. Tiger prawn brings them together into a relationship of mutual dependence. Women catchers, who supposedly cause great harm to biodiversity reserve and embankments, are an important link in the tiger prawn network. Without these women prawn catchers, the network ceases to operate. However, this mutual dependence does not necessarily produce a harmonious community where each player has an equal access to livelihood opportunities or to financial fortunes associated with prawn catching, farming and trading. Among all the key players in the trade, women prawn catchers are the closest to this natural resource because it is they who risk everything to catch prawn seeds in the rivers. But they remain the most deprived. Thus, the network of fishers exists in the form of a hierarchy characterized by varying access to money generated by the trade. The question of how successful one would be in the trade does not merely depend on one's diligence and endurance, but on one's ability to make contact with and manipulate local power structures represented by government functionaries such as irrigation engineers or BSF personnel on the one hand and by panchayat and party leaders on the other. These representatives are certainly not in the thick of the prawn trade, but they, as we have seen, play a crucial role in sustaining prawn farming in the region. There occur constant negotiations between these dealers/traders and people who are wielders of power at the local level, and I have documented some of these in this chapter.

In all of this, the state that we come across appears more disaggregated, the one that is marked by a discontinuity in policies and practices rather than continuity. While, on the one hand, the state comes down heavily on women prawn catchers for their alleged role in environmental degradation and involves NGOs and civic bodies in an attempt to sensitize people to the negative impact of prawn catching, on the other, the state is found deeply implicated in local prawn politics. In the opinion of the policymaking state, prawn catching along the riverbanks may not be sustainable. However, prawn farming and trading is sustained by frequent embankment breaking, bheri making, cross-border trading, all of which are expressions of the state's complicity in local prawn politics. Prawn politics enables us to go beyond the binaries of sustainability and unsustainability. In the above processes, state functionaries such as irrigation engineers, BSF personnel, panchayat and party leaders are found involved at various levels. Their involvement problematizes the idea of a trans-local discrete state, articulating environmental policies. For long, the state has been viewed from the point of view of the dichotomous model of state–society relations (Nugent 1994). A consequence of this kind of thinking is the taken-for-granted perception of the state as some kind of centralized and unified entity. Abrams suggests that the state is not the reality that stands behind the mask of political practice, but it is itself the mask that prevents our seeing political practice as it is (Abrams 1988, 58). Political parties and panchayat leaders engaged in fierce struggles over the control of bheris are certainly not to be viewed as discrete entities at work in the so-called civil society. These struggles and clashes often resulting in the killing of party cadres are important pointers to how the state power is organized and deployed bottoms up rather than top down.

Notes

[1] Tiger prawn and bagda will be used interchangeably in this chapter.

[2] This chapter uses the term prawn politics interchangeably with prawn trade, prawn catching or farming. The term is used to refer to certain activities such as embankment breaking, fishery making or cross-border selling which are directly related to the trade. By prawn politics, I also refer to the activities of people – party leaders or panchayat politicians – who are not directly related to prawn trade, but they make prawn an indispensable part of the power structure as it unfolds in the Sundarbans.

[3] I am indebted to Jalais for her discussion on tiger prawn, the Blue Revolution (2010, 123) in the poverty-stricken Sundarbans. However, I remain entirely responsible for the structure of arguments as it unfolds in the chapter on tiger prawn trade and politics.

[4] Hein makes a mention of the Supreme Court judgment in his paper on shrimp farming along the eastern coast of India. According to Hein, as a result of the Supreme Court judgment and the Aquaculture Bill there are debates going on between the Ministry of Agriculture, Ministry of Environment and Forest, shrimp farming industry and various environmental and social NGOs. For details see Hein (2000) and for the details of the court judgment also see Sharma (1998) at http://base.d-p-h. info/en/fiches/premierdph/fiche-premierdph-4040.html.

[5] Sarkar and Bhattacharya argue that the estuarine delta offers an excellent nursery for most of the brackishwater finfish and shellfish. Women in their desperate bid to separate prawn seeds throw away a major portion of their haul, thereby wasting juveniles of economic and uneconomic varieties of finfish and shellfish. For further details see Sarkar and Bhattacharya (2003).

[6] Fishery and bheri are used interchangeably in the chapter. However, the English word fishery has found its way into the colloquial world of the villagers. Villagers very often use the word fishery as part of their everyday Bengali language.

[7] After five years, the lease normally gets renewed.

[8] Such huge land-holding at the disposal of a single farmer is difficult to come by in view of the land reform policies adopted by the Left-front government which imposed ceiling on individual land-holding.

[9] The TMC had formed the board in the South 24 Parganas' Zilla Parishad having captured fifty-five out of eighty-one seats in the entire district. This shows the growing political influence and the electoral presence of the party.

[10] For the purposes of confidentiality, the name of the island where this incident took place has been deliberately withheld. Pintu is a fictitious name geven to the deceased.

7

Conclusion

Eroding embankments, state machinery and vulnerable lives

The Sundarbans islanders do not have proper roads, bridges across the rivers or electricity in the villages. In these respects, they remain as deprived as those living in forests elsewhere. The Sundarbans delta is pregnant with possibilities. Therefore, to assign primacy to the needs of people, fresh developmental thinking needs to be generated in the region. However, thinking needs to be translated fast into practice, for otherwise, one will not be surprised if tidal waves with their surging water level destroy weak embankments inundating all the fifty-four inhabited islands[in the Indian Sundarbans] ... One has to pay huge cost, if the Sundarbans – a rich resource site, a unique confluence of water, forests and settlements, one of the largest mangrove regions, an elixir to the tired nerves of Kolkata, West Bengal and perhaps India – continues to remain neglected. To ensure that the Sundarbans survives, there need to be concerted efforts on the part of the panchayats, voluntary agencies, administration of the two districts, the state and central governments towards building and conserving the Sundarbans embankments. If the embankments are built and properly taken care of, all other developments in the region will be sustainable. But if the embankments are neglected and a priority is given to piece-meal development, the Sundarbans will eventually relapse into uninhabited jungles. Protecting the embankments not only means preserving the lives of 43 lakhs [4.3 million] people of the Sundarbans, but also means achieving a delicate balance between people living in the Sundarbans and those living outside (Nayek 2005, 2; translated from vernacular).

The embankments are found still lying breached even five years after the cyclone Aila. People's vulnerability to disaster continues unabated

(Oliver-Smith 1996, 315) as land continues to erode and people are found to be on the retreat. Hence the question remains is where will all this backtracking lead? In this book, our journey had started with the eroding lives of north Kusumpur where subsequently the cyclone Aila had caused many to leave their home. For those still living in this narrow stretch, the land had become even narrower. As has been discussed in the earlier chapters of the book, people keep negotiating their time with the rivers and with the government machinery meant for flood control.

One is struck by the absence of a disaster management policy on the part of the government. Instead, what we witness are discrete flood control or embankment building practices based on the implicit assumption that development interventions at the end of the day are unsustainable in the region. It is as if frequency and inevitability of storm, tidal waves or land erosion point to the futility of human settlement in the delta. This is the assumption that informs the shaping of policy discourses in the corridors of power. Policy inaction is often justified through recourse to an argument that the Sundarbans is an instance of premature land settlement where people settled much before the land formation or elevation was complete. The embankments built in defiance of the course of rivers to prevent the flooding of low-lying settled lands are erosion prone anyway and therefore unsustainable. This is the broad view that permeates piecemeal activities of the government department meant exclusively for the development of the Sundarbans region, namely Sundarban Affairs Department. We all know how the settlement history in the Sundarbans had largely been part of the colonial regime of revenue and commerce. It is as if the clock of history needs to be turned back to rectify this primal mistake the early settlers committed, or else people living on these islands in the present day Sundarbans might as well be prepared to pay the price for the mistake their forefathers committed when they reclaimed and settled on these lands.

In the irrigation department's scheme of things, big dams appear as experiments in science and monuments of national heritage and the Sundarbans embankments are of marginal significance. A considerable importance is attached to big dam projects in West Bengal, while emergency stopgap measures are reserved for the Sundarbans. It is interesting to note how the post-independent Indian governments, right, left or centre, inherit the colonial legacy of building dams and setting up sophisticated models of environment control. Irrespective of whether they succeed or fail, big dams conceived in painstaking mathematical calculations remain the centrepiece of post-independent India's narrative of national development. By contrast, the Sundarbans embankments, the fragile mud walls, are not worthy of attention by the community of engineers and specialists. Even after five years since the cyclone, the broken and breached embankments bear testimony to governmental neglect and apathy. The Sundarbans' local newspaper *Badweep Barta* states,

> What did the irrigation department of the West Bengal government do in taking care of the embankments in thirty-two years? After the Aila, the left government of West Bengal has planned to spend Rs. 10,000 crores in building and constructing the Sundarbans embankments. However, this hardly boosts the morale of the poverty and Aila-struck islanders. In 2006, Rs. 85 lakhs was spent rebuilding 500 meters of the embankment in Basanti, the embankment collapsed in fifteen days (*Badweep Barta* 2009c; translated from vernacular).

The report is significant in many ways. It accuses the government and the irrigation department of not having done anything to strengthen the Sundarbans embankments in the last thirty-two years, the period of left rule in West Bengal. It also draws attention to the futility of piecemeal flood control measures such as the one undertaken in Basanti block in 2006. The report also suggests that Rs. 85 lakhs worth embankment did not last beyond fifteen days. In other words, the report points to the irrigation department's malpractices in flood control measures.

Building the Sundarbans embankments, no matter how unsustainable they are, presents the engineers and contractors with the possibility of making money. This engineer–contractor network often works in close tandem with local party and panchayat leaders. To ensure that engineers and contractors have the weight of the political parties and local politicians behind them, local party leaders are drawn into this network and money earned from stopgap flood control measures are believed to be shared amongst different people located in the network. The embankments could be of marginal significance, but the expenses planned for building embankments are always inflated because profit is perceived as travelling all the way from functionaries stationed at the local-level to the upper-level bureaucrats in the irrigation department. This book documents how this development disempowers and produces a sense of vulnerability among people. While studying the reconstruction process in post-disaster Turkey, D'Souza[1] argues that development initiatives could be seen as attempts to reduce people's vulnerability to disasters (D'souza 1986, 49). However, in the Sundarbans, people's vulnerability to disaster is induced by development interventions that are not only profit making, but land grabbing as well. As we have seen earlier in the book, the bigger the land acquisition for embankment building, the larger is the share of profit.

Islanders' vulnerability clubs them into a collective, as we have seen this in the case of the villagers of north Kusumpur who constantly backtracked as they lost land to the flood control machinery of the government. The book has shown how displacement and governmental apathy forged villagers of Garantala into a collective that resisted governmental inaction by forcibly occupying a vacant land on Garantala island itself. The community of the vulnerable and marginalized have the potential to challenge the gigantic machinery of the government and make it recognize their claims. However, such robust outbursts against governmental apathy and inaction are not easy to come by. This is because in a land, where nothing settles for good, where people struggle to stay afloat, individuals look for some livelihood opportunities in the form of a breather. It is their search for livelihood or what Lahiri-Dutt and Samanta call banchbar upay (ways and means to survive; Lahiri-Dutt and Samanta 2014, 159) that brings them into a relation with the government machinery which otherwise marginalizes and disempowers them.

Thus, we come across construction workers, dafadars and beldars' increasing negotiations with the irrigation machinery with its elaborate apparatus of flood control. The workers and dafadars belong to the vulnerable collective of north Kusumpur, yet they compete and at times outwit each other in trying to grab the resources that go with construction or flood control projects funded by the irrigation department. When I visited north Kusumpur a year after the cyclone, I found many houses damaged and still not repaired. It was true that the villagers were economically constrained and therefore, understandably it was difficult for them to rebuild their houses. However, a closer scrutiny revealed that villagers of north Kusumpur, at least a majority of them, did not deliberately rebuild their houses because they were waiting for government money. We already know from our discussion in the earlier chapters (Chapter 4 in particular) that immediately after the cyclone, government declared compensation for fully and partially damaged houses.

However, when I visited Kusumpur after a year the money still had not reached the people. The villagers of north Kusumpur were given to understand that a team of government officials would visit different parts of the Sundarbans to inspect the condition of houses before the final release of funds. On hearing this, the villagers decided not to build their houses lest they run the risk of losing the compensation money. During the period after the cyclone when embankments were constructed and repaired on an emergency basis in the

Sundarbans, it was Dinu who earned enough by being the irrigation department's labour supervisor at many such construction sites. With the help of his dafadar's job, he could build his house in north Kusumpur. Thus, Dinu's rebuilt and done-up house stood out among the damaged and broken houses as though it was a mark of his new found economic solvency. One does not know when Dinu's rebuilt house would fall prey to the tidal erosion or irrigation department's land acquisition. But at least temporarily Dinu found a livelihood in his otherwise uncertain life. Despite being part of the vulnerable landscape and equally exposed to the risk of flooding and displacement, the adivasi beldars emerge as the agents of the state occupying the lowest rung of the irrigation machinery. As irrigation functionaries, they represented the interests of the government department even when such interests ran contrary to their interests as the Sundarbans islanders who, like their neighbours, remained vulnerable to land acquisition and land loss. They were also found to be in constant negotiations with the government for their rights and entitlements as irrigation employees. In the face of these everyday negotiations and strategies of survival, our neatly constructed categories such as state, people, community and indigeneity all crumble into an untidy jumble of contradictions (Baviskar 1997, vii).

Contradictions loom large when we turn to tiger prawn catching, farming and trading in the region. It is once again people's search for a livelihood, in a region where agriculture is uncertain because of continuous saline ingress, land erosion and virtually non-existent winter cultivation, that helps explain why and how prawn trade flourishes in the Sundarbans. The book lays bare a wide network of interests around prawn catching and trade. Women from poor families catch tiger prawn seeds along the riverbanks and run the risk of being killed by sharks or crocodiles. Prawn trade flourishes because these women are supported by many other players in this prawn network. We come across islanders engaged in divergent, disparate and often self-destructive acts, which lend visibility to prawn trade. The protective mud walls of embankments are deliberately tampered with or broken to release saline water in agricultural fields so as to turn them into prawn fisheries. People who look upon embankments as their very lifeline do not hesitate to destroy them in search of a livelihood. If building embankments presents people with a livelihood (as shown in the case of a dafadar), breaking them also generates possibilities of livelihoods. Prawn dealers like Niranjan were found striking deals with local political leaderships to survive in the trade and people formed close networks amongst themselves to resort to cross-border prawn trade in Bangladesh.

Tiger prawn leads us to prawn society where fisheries or bheris flourish under local political patronage; fisheries are means through which parties expand their sphere of political influence; fisheries are pivots around which intra-party and inter-party conflicts intensify, resulting in the killing of party cadres. I have focused on the complicity of the local party and panchayat leadership, the irrigation functionaries, border security personnel and the villagers as prawn catchers, dealers, sellers, fishery owners and party cadres cum traders (when fishery owner happens to be a local person) in prawn politics. And in all of these people appear as much a disaggregated category as the state. The state looks upon prawn catching as having negative impact on the ecology of the Sundarbans. The state departments (such as forest or Sundarban Affairs) single out women prawn catchers as being responsible for depleting biodiversity and weakening embankments. However, the state, which portrays prawn catching as a destructive livelihood and comes down on women prawn catchers for their alleged role in environmental degradation, is found itself implicated in prawn trade and prawn politics. The state appears disaggregated marked by a discontinuity between the state's stand against tiger prawn catching and the state's complicity in local prawn trade.

Can we still think of a disaster management policy? Need for a concerted effort

However, the question that remains is where do we go from here? Can we still think of a disaster management policy by the state? This question is important in the face of the portrayal of state, society and people that I have attempted above. It is important to recall Nayek's observation with which this concluding chapter has opened. Nayek argues that there should be a concerted effort on the part of panchayat, district administration, state and central governments towards protecting and preserving the Sundarbans embankments. The phrase 'concerted effort' is important in the context of the Sundarbans because there has never been any concerted effort on the part of the government towards ameliorating the vulnerable condition of the islanders. The government departments have always acted in isolation. While the irrigation department acquires land and builds embankments, the SAD only duplicates the efforts of departments such as public works or agriculture. Rarely, if ever, had the SAD collaborated with the irrigation department in initiating development which would secure people's confidence in governmental power.

The emphasis here is on the concerted effort of the government departments in evolving an effective disaster management policy. Policies are technologies of governance (Shore and Wright 1997, 29) that tend to construct their subjects as objects of power (Ibid, 3).[2] But policies also have the potential to assign a sense of agency and purpose to their subjects. Here, the focus is not merely on protecting and preserving the Sundarbans embankments because embankments no matter how strongly built are bound to be eroded by river currents. We are not concerned with land acquisition and temporary and emergent embankment repair. A comprehensive disaster policy means efforts on the part of the departments like Sundarban Affairs and Irrigation to establish information bank on river currents and erosion. In doing this, they should collaborate with organizations such as the River Research Institute (RRI). The irrigation department is only interested in dams but not in rivers. The database and insight of the RRI can be used to set up an information bank on the rivers in the Sundarbans. This information bank can be used to prevent actual or potential displacement of people. Land acquisition can be justified only when it is accompanied by a definite policy of relocation with compensation, a relocation package that reaches different sections of the populace. In working out compensation packages, the irrigation and the SAD should act in consultation with the land assessment department of the West Bengal government. Immediately after the Aila, the government declared compensation packages for land acquisition, but this was conveniently shelved and forgotten after the crisis was over. A disaster management policy such as the one mentioned above will not instantly provide a solution to the manifold problems the islanders face in the region. Still, some initiative in the direction of relocation and compensation would certainly prevent people's further vulnerability and marginalization. All this while the islanders in the Sundarbans have lost lands, but rarely, if ever, have they been compensated for their lost lands.

Rethinking the Sundarbans development: Beyond the exclusivist conservation paradigm

However, at another level a democratic restructuring of what is meant by Sundarbans development seems necessary. A disaster management policy could materialise successfully only if there is a shift in the larger perception of what it means to undertake development of the region. People's needs and desires are often found subsumed to the dominant needs of the Sundarbans conservation and it is difficult to salvage a notion of people's Sundarbans from the more hegemonic representation of the delta as a wildlife sanctuary. The Marichjhanpi event discussed in the second chapter serves to remind us of the centrality of this image. The image had been so central and powerful that the left government denied the refugees their right to settlement on a forested island. Ecology has never been the centrepiece of communist movement in Bengal. But when the communists turned to the Sundarbans they placed primacy on ecology and denied poor peoples' right to land.

The image of the region as a forested landscape and an abode of wildlife is made to appear as though it is a natural fact. My book has interrogated the construction of the dominant image of the Sundarbans and shown how this image had been constructed by technologies of colonial rule, most notably the colonial gazetteer writing and by a wide variety of literature produced on the Sundarbans in post-independent India. The sustainable development of the region appears to require that people should be excluded from the development process, because their needs are perceived as obstacles to the successful conservation of the wildlife habitat. This is the development imperative that has informed policy formulations of the government and powerfully shaped the orientation of the bureaucrats and state functionaries charged with the Sundarbans' conservation. It is this grand vision of conservation that restricts people's livelihood needs and access to resources. In fact, the increasing population of the Sundarbans is viewed as being responsible for the erosion of biodiversity and ecological degradation of the region. We have also seen how the state portrays prawn seed catching by women as a dangerous livelihood instrumental in causing biodiversity loss. One wonders how the state could restrict and control women prawn catchers when tiger prawn trade flourishes as the most sought-after livelihood throughout the Sundarbans. Thus, the state's approach to the Sundarbans development is lopsided and has injustice built into it (Mukhopadhyay 2009, 121).

The grand vision of conservation described above leads us to a dominant model of conservation, namely the displacement model. For conservation to be effective, local population and their needs have to be kept at bay. The local population may be relocated away from the conservation site so as to ensure the smooth functioning of a national park or wildlife sanctuary. Sharma and Kabra (2007) and Johari (2007) provide a critique of this model of conservation. My book is not about conservation. Still the book has focused on an adivasi bauley, a tiger charmer and forest goer and the story of his journey into the forests (see Chapter 5). I have presented the bauley's narrative to show his sense of responsibility and commitment towards his profession as a forest goer. As a worshipper of Bonbibi, he was deeply respectful towards the existence of non-humans in the forest and their rights to forest resources. For him, the sanctity of the forest was of crucial importance and under no circumstances could he afford to violate the forest which remained a source of his livelihood. Thus, a bauley's narrative of forest expedition together with his care and concern for the forest do give us a sense of how people in the Sundarbans emerge as conservers of forest.

Jalais' book *Forest of Tigers* (2010)[3] is a telling commentary on people's perspective on conservation. Her book concerns the Sundarbans forests reflecting on the ways people conceive of forests, tiger and the relation they have with each other. According to Jalais, people's perceptions of forest and tiger are significant because such perceptions resist a more dominant and powerful model of statist conservation which views people merely as obstacles. Over the years, there has occurred a shift from conventional exclusive models of conservation to more inclusive people-friendly paradigms which stand at the interface of sociological and biological expertise (Rangarajan and Shahabuddin 2007). Sociologists and conservation biologists constantly exchange their perspectives and insights in evolving this participatory and inclusive paradigm of conservation (Baviskar 2005; Rai 2007; Das 2007; Datta 2007).[4] Based on pragmatic considerations, these conservation paradigms combine a more nuanced view of ecological processes with an appreciation of the importance of the long-standing livelihood rights of human populations (Baviskar 2005, 294). The institutional relationships between projects, wildlife institutions and communities are fundamental to promoting participation and sustainability (Rodgers et al. 2005, 355).

Local people's engagement with the forest or resource site is often mediated by livelihood needs. Therefore, experiences born of such engagement can prove to be vital inputs in the conservation programme. India is already a witness to many such conservation programmes based on people-friendly paradigms focusing on the coexistence of wildlife and local population. The conservation paradigm pursued in the Sundarbans so far was based on the principle of exclusion of people. However, the conservation paradigm needs to be based on a recognition of people as the legitimate occupants of land and not merely as 'tiger food' (Jalais 2010, 11). Inhabited islands are as much a part of the Sundarbans as the reserved forests. All this while, the Sundarbans as an abode of humans lay hidden under a dominant and overarching image of the region as a natural wilderness and a forested tiger land. It is thus necessary to disentangle people's own needs and priorities from the grand narrative of the Sundarbans conservation and to allow the former to inform the policies and initiatives of the government. What I have considered above are a few suggestions, by no means exhaustive or conclusive, for changes in approaches to the Sundarbans development. Some initiatives along the above lines may demonstrate a clearer recognition on the part of the state that people will continue living in the Sundarbans and their needs cannot remain subservient to those of wildlife.

Notes

[1] D'Souza's work focuses on reconstruction process in four villages of western Turkey following the Gediz earthquake. The study is mainly concerned with what recovery meant for different affected communities, what role the government played in making recovery available to different sections of people. Far from viewing government initiatives as attempts to reduce people's vulnerability to disasters, the study viewed reconstruction efforts by the government as further inducing and aggravating the vulnerability of people. D'Souza argues that disasters can accentuate and accelerate the trends which were apparent in a given society prior to the disaster. The communities which were already solvent before the disaster gained substantially from the post-earthquake reconstruction process while those who were economically weaker before the earthquake remained vulnerable. For details, see D'Souza (1986).

[2] Shore and Wright's (1997) anthropology of policy offers a fresh perspective on policy. They look upon policy as a vehicle for political legitimacy. Attempts are made to establish relation between policy and morality. By thinking of policy as a cultural category and political technology, a new impetus has been given to political anthropology. They also understand policy as a linguistic category, thereby establishing the connection between discourse and power. Policy has also been understood as governmentality. Building on Foucault, the authors see policy as a site for constant negotiations between the governed and those in governance. Policy is also viewed as a cultural agent constructing national identity.

[3] Jalais' work deals with forest, tiger and people, focusing on the ways they establish relations amongst each other. With tiger, people share a harsh geographical space and unique history of displacement. By talking about forest, tiger and Bonbibi, the forest deity, people articulate their locations in the fluid and transient world of the Sundarbans. For details, see Jalais (2010).

[4] A host of new approaches to conservation have surfaced. They are as follows:

Baviskar examines the question of people's involvement in resource conservation. Baviskar highlights problems associated with the ecodevelopment project in the Great Himalayan National Park in Himachal Pradesh. According to the author, the practice of ecodevelopment as it unfolds in Great Himalayan National Park has stifled the possibility of dialogical outcomes. Baviskar argues in favour of a more flexible approach to conservation where scientists, foresters, villagers and other stakeholders can jointly set the terms of reference. For further details see Baviskar (2005, 267–99).

Das explores the theme of participatory conservation in Kailadevi Wildlife Sanctuary in Rajasthan. She shows how caste and community-based cleavages among Kailadevi's residents inform structure and functioning of collective institutions. Das argues in favour of people's involvement in conservation, but also suggests that participatory paradigm needs to be analysed in the broader historical and political context of conservation keeping in view cross cutting ties of various agents involved. See Das (2007, 113–46).

Rai's work on Uttara Kannada in southern India argues in favour of granting control of forest to forest dependent people, because continuous human use of forest plays a crucial role in regenerating the forest landscape. He argues that local community-based conservation institutions, flexible and adapted to local contexts, should be the building blocks of a much larger participatory framework. See Rai (2007, 147–62).

Datta's work on Namdapha Tiger Reserve in Arunachal Pradesh focuses on dialogue with the members of the Lisu tribe (tribe of hunters who inflicted damage on the tiger reserve), drawing them into the modern conservation process by way of creating alternative livelihoods for the tribe. For details see Datta (2007, 165–209).

Bibliography

Abrams, P. 1988. 'Notes on the Difficulty of Studying the State'. *Journal of Historical Sociology* 1 (1): 58–89.

Agarwal, B. 1986. 'Women, Poverty and Agricultural Growth in India'. *The Journal of Peasant Studies* 13 (4): 165–220.

———. 1988. 'Neither Sustenance nor Sustainability: Agricultural Strategies, Ecological Degradation and Indian Women in Poverty'. In *Structures of Patriarchy: The State, the Community and the Household*, edited by B. Agarwal. London: Zed Books.

———. 1989. 'Women, Land and Ideology in India'. In *Women, Poverty and Ideology in Asia: Contradictory Pressures, Uneasy Resolutions*, edited by H. Afshar and B. Agarwal. London: Macmillan.

———. 1992. 'The Gender and Environment Debate: Lessons From India'. *Feminist Studies* 18 (1): 119–58.

———. 1994. *A Field of One's Own: Gender and Land Rights in South Asia*. Cambridge: Cambridge University Press.

Agrawal, A. 1999. 'Community-in-Conservation: Tracing the Outlines of an Enchanting Concept'. In *A New Moral Economy for India's Forests: Discourses of Community and Participation*, edited by R. Jeffery and N. Sundar. New Delhi, London: Sage Publications.

Ajker Basundhara. 2010. 'Bagda Min Dharai Bhangchhe Nadibund Luptoprai Matsakul Kobe Bandho Hobe Ei Atmaghati Pesha' (Catching Tiger Prawn Seeds has Destroyed Embankments, Caused Depletion of Fishing Resources. When will this Self-Destructive Profession Come to an End). *Aajker Basundhara*, Basanti, September 1.

Alvares, C. 1988. 'Science, Colonialism and Violence: A Luddite View'. In *Science, Hegemony and Violence: A Requiem for Modernity*, edited by A. Nandy. Delhi: Oxford University Press.

Anderson, B. 1991 [1983]. *Imagined Communities: Reflections on the Origin and Spread of Nationalism*. London: Verso.

Aranyadoot. 2009a. 'Basanti Block-e Trinamul-er Shakti Briddhi' (Rise in Trinamul's Strength in Basanti Block). *Aranyadoot*, Sonarpur, July 1–31.

———. 2009b. 'Gosaba-ei CPI-M er Panchayat Sadasya ra Sadalbal e Trinamul-e' (The CPI-M Panchayat Members of Gosaba En Masse Join Trinamul). *Aranyadoot*, Sonarpur, July 1–31.

———. 2009c. 'Loksabha Nirbachan 2009-er Falafal' (Results of Parliamentary Elections of 2009). *Aranyadoot*, Sonarpur, August 1–14.

Ascoli, F. D. 1921. *A Revenue History of the Sundarbans: From 1870 to 1920*. Calcutta: Bengal Secretariat Book Depot.

Badweep Barta. 2003. 'Be-Aini Bheri-r Birudhhe Sarab CPI-M' (CPI-M is Vocal Against Illegal Fishery). *Badweep Barta*, Canning, March 1–15.

———. 2007. 'Chingri Anyo Macher Aakal Deke Enechhe' (Prawn has Caused Depletion of Other Fish Species). *Badweep Barta*, Canning, February 16–28.

———. 2009a. 'Aila-e Durgato Sundarban: Tran-e Dalbaji' (Aila-Stricken Sundarbans: Partisan Vested Interests Over Relief). *Badweep Barta*, Canning, June 1–30.

———. 2009b. 'Masjidbati te Aila-e Tran Niye Rajniti' (Politics Over the Aila Relief Distribution in Masjidbati). *Badweep Barta*, Canning, August 16–31.

———. 2009c. 'Sammekshar Dayitwa Kar? Battris Bachharer Sasane Bund Samanya Unchu Holo Na Keno' (Whose Responsibility was the Survey? Why wasn't the Height of the Embankment Slightly Raised in Thirty-Two Years). *Badweep Barta*, Canning, August 1–15.

———. 2012. 'Min Babsayee Khun' (Seed Trader Killed). *Badweep Barta*, Canning, January 1–15.

Bandyopadhyay, D. 1989. *Budget Speech of the Minister-in-charge of Irrigation and Waterways Department 1989–90*. Calcutta: Government of West Bengal.

———. 1990. *Budget Speech of the Minister-in-charge of Irrigation and Waterways Department 1990–91*. Calcutta: Government of West Bengal.

———. 1993. *Budget Speech of the Minister-in-charge of Irrigation and Waterways Department 1993–94*. Calcutta: Government of West Bengal.

————. 1994. *Budget Speech of the Minister-in-charge of Irrigation and Waterways Department 1994–95*. Calcutta: Government of West Bengal.

————. 1998. *Budget Speech of the Minister-in-charge of Irrigation and Waterways Department 1998–99*. Calcutta: Government of West Bengal.

————.1999.*Budget Speech of the Minister-in-charge of Irrigation and Waterways Department 1999–2000*. Calcutta: Government of West Bengal.

————.2000.*Budget Speech of the Minister-in-charge of Irrigation and Waterways Department 2000–2001*. Calcutta: Government of West Bengal.

Banerjee, A. 1998. *Environment, Population and Human Settlements in Sundarban Delta*. New Delhi: Concept Publishing Company.

Banerjee, P. 2010. *Borders, Histories, Existences: Gender and Beyond*. New Delhi: Sage Publications.

Banerjee, P., S. Basu Ray Chaudhury, S. Kumar Das. 2005. *Internal Displacement in South Asia: The Relevance of the UN's Guiding Principles*. New Delhi: Sage Publications.

Basu, J. 2000. *Bamfront Sarkarer Teyish Bachar* (Twenty-Three Years of the Left-front Government). West Bengal: Information and Culture Department, Government of West Bengal.

Basu Ray Chaudhury, A. 2011. 'Narrated Time and Constructed Space: Remembering the Communal Violence of 1950 in Hooghly'. In *Women in Indian Borderlands*, edited by P. Banerjee and A. Basu Ray Chaudhury. New Delhi: Sage.

Bates, C. and M. Carter. 1994. 'Tribal Migration in India and Beyond'. In *The World of the Rural Labourer in Colonial India*, edited by G. Prakash. Delhi: Oxford University Press.

Baviskar, A. 1997. *In the Belly of the River: Tribal Conflicts over Development in the Narmada Valley*. New Delhi: Oxford University Press.

————. 2005. 'States, Communities and Conservation: The Practice of Ecodevelopment in the Great Himalayan National Park'. In *Battles Over Nature: Science and the Politics of Conservation*, edited by V. Saberwal and M. Rangarajan. Delhi: Permanent Black.

Bayly, C. A. 1996. *Empire and Information: Intelligence Gathering and Social Communication in India, 1780–1870*. Cambridge: Cambridge University Press.

Bhunia, M. S. 2011. *Budget Speech of the Minister-in-charge of Irrigation and Waterways Department 2011–12*. Kolkata: Government of West Bengal.

————. 2012. *Budget Speech of the Minister-in-charge of Irrigation and Waterways Department 2012–13*. Kolkata: Government of West Bengal.

Breman, J. 2000. "'I am the Government Labour Officer …:" State Protection for the Rural Proletariat of South Gujarat'. In *Politics and the State in India*, edited by Z. Hassan. Delhi: Sage Publications.

Breman, J. and E.V. Daniel. 1992. 'Conclusion: The Making of a Coolie'. *The Journal of Peasant Studies* 19 (3&4): 268–95.

Brow, J. 1988. 'In Pursuit of Hegemony: Representations of Authority and Justice in a Sri Lankan Village'. *American Ethnologist* 15 (2): 311–27.

Chambers, R. 1983. *Rural Development: Putting the Last First*. England: Pearson Education Limited.

———. 1994. 'The Origins and Practice of Participatory Rural Appraisal'. *World Development* 22 (7): 953–69.

———. 1995. 'Paradigm Shifts and the Practice of Participatory Research and Development'. In *Power and Participatory Development: Theory and Practice*, edited by N. Nelson and S. Wright. London: Intermediate Technology Publications.

———. 1997. *Whose Reality Counts?: Putting the First Last*. London: ITDG Publishing.

Chatterjee, N. 1992. *Midnight's Unwanted Children: East Bengali Refugees and the Politics of Rehabilitation*. PhD diss., USA: Brown University.

Chatterjee, P. 1982. 'Agrarian Structure in Pre-Partitioned Bengal'. In *Perspectives in Social Sciences: Three Studies on the Agrarian Structure in Bengal 1850–1947*, edited by A. Sen, P. Chatterjee and S. Mukherjee. Calcutta: Centre for Studies in Social Sciences.

———. 1995. *The Nation and Its Fragments: Colonial and Postcolonial Histories*. Delhi: Oxford University Press.

———. 1998. 'Community in the East'. *Economic and Political Weekly* 33 (6): 277–82.

———. 2004. *The Politics of the Governed: Reflections on Popular Politics in Most of the World*. Delhi: Permanent Black.

Chattopadhyay, H. 1999. *The Mystery of the Sundarbans*. Calcutta: A. Mukherjee and Co. Private Limited.

Chattopadhyay, S. S. 1994. *'Foreword'*, *Annual Administration Report 1981–82 to 1991–92*, Department of Sundarban Affairs. Calcutta: Government of West Bengal.

Chattopadhyay, S. S. 2009. 'Gone with the Wind: Cyclone Aila Leaves the People of the Sunderbans Wallowing in Misery'. *Frontline*, July 3. 26 (13): 32–36.

Chaudhuri, K. 2002. 'The Vanishing Tiger'. In *The Statesman*. October 13.

Choudhury, N. C. and S. K. Bhowmik. 1986. 'Migration of Chota Nagpur Tribals to West Bengal'. In *Studies in Migration: Internal and International Migration in India*, edited by M. S. A. Rao. New Delhi: Manohar.

Cohn, B. S. 1997. *Colonialism and Its Forms of Knowledge: The British in India*. Delhi: Oxford University Press.

Damodaran, V. 2005. 'Indigenous Forests: Rights, Discourses and Resistance in Chotanagpur, 1860–2002'. In *Ecological Nationalisms: Nature, Livelihoods, and Identities in South Asia*, edited by G. Cederlöf and K. Sivaramakrishnan. Delhi: Permanent Black.

Das, A. K., S. Mukherji, M. K. Chowdhuri. 1981. *A Focus on Sundarban*. Calcutta: Editions Indian.

Das, P. 2007. 'The Politics of Participatory Conservation: The Case of Kailadevi Wildlife Sanctuary, Rajasthan'. In *Making Conservation Work: Securing Biodiversity in this New Century*, edited by G. Shahabuddin and M. Rangarajan. Ranikhet Cantt: Permanent Black.

Das, S. K. 2005. 'India: Homelessness at Home'. In *Internal Displacement in South Asia: The Relevance of the UN's Guiding Principles*, edited by P. Banerjee, S. Basu Ray Chaudhury, and S. K. Das. New Delhi: Sage.

Datta, A. 2007. 'Threatened Forests, Forgotten People'. In *Making Conservation Work: Securing Biodiversity in this New Century*, edited by G. Shahabuddin and M. Rangarajan. Ranikhet Cantt: Permanent Black.

De, R. 1990. *The Sundarbans*. Delhi: Oxford University Press.

Department of Agriculture and Horticulture. n.d. *Reconstruction and Rehabilitation Reports of Agriculture/Horticulture Departments of South 24 Parganas*. West Bengal: Government of West Bengal.

Devi, M. 1977. *Aranyer Adhikar* (Right to Forest). Kolkata: Karuna Prakashani.

Directorate of Forests. n.d. *Sundarban Biosphere Reserve*. Calcutta: Government of West Bengal.

———. 2004. *Indian Sundarbans: An Overview*. *Sundarban Bio-sphere Reserve*. Calcutta: Government of West Bengal.

Dirks, N. B. 1997. 'Foreword'. In *Colonialism and Its Forms of Knowledge: The British in India*, by B. S. Cohn. Delhi: Oxford University Press.

———. 2001. *Castes of Mind: Colonialism and the Making of Modern India*. Princeton: Princeton University Press.

D'Souza, F. 1986. 'Recovery Following the Gediz Earthquake: A Study of Four Villages in Western Turkey'. *Disasters* 10 (1): 35–52.

D'Souza, R. 2006. *Drowned and Dammed: Colonial Capitalism and Flood Control in Eastern India*. New Delhi: Oxford University Press.

Duyker, E. 1987. *Tribal Guerrillas: The Santals of West Bengal and the Naxalite Movements*. Delhi: Oxford University Press.

Eaton, R. 1987. 'Human Settlements and Colonization in the Sundarbans 1200–1750'. Paper presented at the workshop on *The Commons in South Asia: Societal Pressures and Environmental Integrity in the Sundarbans*, A workshop held at Smithsonian Institution, Washington, DC. November 20–21. www.smartoffice.com/Tiger/Eaton.html. Accessed September 2002.

Escobar, A. 1995. *Encountering Development: The Making and Unmaking of the Third World*. Princeton, NJ: Princeton University Press.

Ferguson, J. 1990. *The Anti-Politics Machine: "Development," Depoliticisation and Bureaucratic Power in Lesotho*. Cambridge: Cambridge University Press.

Fuller, C. J. and V. Bénéï, eds. 2001. *The Everyday State and Society in Modern India*. London: Hurst and Company.

Gadgil, M. and R. Guha. 1992. *This Fissured Land: An Ecological History of India*. Delhi: Oxford University Press.

Gardner, K. 1999. 'Location and Relocation: Home, "the Field" and Anthropological Ethics (Sylhet, Bangladesh)'. In *Being There: Fieldwork in Anthropology*, edited by C. W. Watson. London: Pluto Press.

Ghosh, A. 2013. *Ati Rajniti-r Sankat* (The Dilemma of Hyper Politics). Kolkata: Ebong Mushaera.

Ghosh, A. 2001. *The Shadow Lines*. New Delhi: Ravi Dayal & Permanent Black.

———. 2004. *The Hungry Tide*. New Delhi: Ravi Dayal Publisher.

Ghosh, K. 1999. 'A Market for Aboriginality: Primitivism and Race Classification in the Indentured Labour Market of Colonial India'. In *Subaltern Studies,* vol. 10, edited by G. Bhadra, G. Prakash, and S. Tharu. Delhi: Oxford University Press.

Ghosh, S. 2010. 'Kultali Block-er Deulbari Jogneswarpada-e Aila Bishoy-e Prasnabali Bhitwik Sameeksha' (A Questionnaire-Based Field Survey on Aila at Deulbari Jogneswarpada of Kultali Block), in *Sundarban Basir Sathe* (With the People of the Sundarbans), *Kemon Ache Sundarbaner Manush: Aila Parabarti Ekti Sameeksha* (How the People of the Sundarbans are Doing: A Post-Aila Survey). Kolkata: Manthan Samayiki.

Gillison, G. 1980. 'Images of Nature in Gimi Thought'. In *Nature, Culture and Gender*, edited by C. P. MacCormack and M. Strathern. Cambridge: Cambridge University Press.

Government of India. 2011. *Census 2011: Primary Census Abstract*. Ministry of Home Affairs. Delhi: Office of the Registrar General and Census Commissioner.

Greenough, P. 1998. 'Hunter's Drowned Land: An Environmental Fantasy of the Victorian Sunderbans'. In *Nature and the Orient: The Environmental History of South and Southeast Asia*, edited by R. H. Grove, V. Damodaran, and S. Sangwan. Delhi: Oxford University Press.

Grout, A. 1996. 'Possessing the Earth: Geological Collections, Information and Education in India, 1800–1850'. In *The Transmission of Knowledge in South Asia: Essays on Education, Religion, History, and Politics*, edited by N. Crook. Delhi: Oxford University Press.

Guha, B. and A. Biswas. 1991. 'Lessons From a Study of Some Riverbank Settlements in Sundarban, West Bengal'. *Transactions of the Institute of Indian Geographers* 13 (1): 53–64.

Guha, R. 1989. *The Unquiet Woods: Ecological Change and Peasant Resistance in the Himalaya*. Delhi: Oxford University Press.

———. 1990. 'An Early Environmental Debate: The Making of the 1878 Forest Act'. *The Indian Economic and Social History Review* 27 (1): 65–84.

———. 1998. 'Introduction'. In *Social Ecology*, edited by R. Guha. Delhi: Oxford University Press.

Guha, R. and J. Martinez-Allier. 2000. *Varieties of Environmentalism: Essays North and South*. Delhi: Oxford University Press.

Gupta, A. 1995. 'Blurred Boundaries: The Discourse of Corruption, the Culture of Politics, and the Imagined State'. *American Ethnologist* 22 (2): 375–402.

Gupta, D. 1997. *Rivalry and Brotherhood: Politics in the Life of Farmers in Northern India*. Delhi: Oxford University Press.

Handelman, D. 1981. 'Introduction: The Idea of Bureaucratic Organization'. *Social Analysis* 9: 5–23.

Harris, M. 1988. 'On Charities and NGOs'. In *Putting People First: Voluntary Organisations and Third World Development*, edited by R. Poulton and M. Harris. London: Macmillan.

Haynes, D. and Prakash, G., eds. 1991. *Contesting Power: Resistance and Everyday Social Relations in South Asia*. Berkeley, Los Angeles: University of California Press.

Hein, L. 2000. 'Impact of Shrimp Farming on Mangroves Along India's East Coast'. *Unasylva* 203 51: 48–55.

Hena, A. 2011. *Budget Speech of the Minister-in-charge of Fisheries, Aquaculture, Aquatic Resource and Fishing Harbours Department 2011–2012*. Calcutta: Government of West Bengal.

Herring, R. 1987. 'The Commons and Its "Tragedy" as Analytical Framework: Understanding Environmental Degradation in South Asia'. Paper presented in *The Commons in South Asia: Societal Pressures and Environmental Integrity in the Sundarbans*, the workshop held at Smithsonian Institution, Washington, DC. November 20–21. www.smartoffice.com/Tiger/Herring.html. Accessed September 2002.

Hewitt, K. 1983. 'The Idea of Calamity in a Technocratic Age'. In *Interpretations of Calamity: From the viewpoint of Human Ecology*, edited by K. Hewitt. New York: Allen and Unwin.

Hill, C. V. 1990. 'Water and Power: Riparian Legislation and Agrarian Control in Colonial Bengal'. *Environmental History Review*14 (4): 1–20.

———. 1991. 'Philosophy and Reality in Riparian South Asia: British Famine Policy and Migration in Colonial North India'. *Modern Asian Studies* 25 (2): 263–79.

Hindustan Times. 2002. "Sundarbans Among 37 Pristine Areas 'Critical to Earth's Survival'". *Hindustan Times*. December 5.

Hobart, M. 1993. 'Introduction: The Growth of Ignorance?'. In *Anthropological Critique of Development: The Growth of Ignorance*, edited by M. Hobart. London: Routledge.

Hunter, W. W. 1998 [1875]. *A Statistical Account of Bengal Vol 1: Districts of 24 Parganas and Sundarbans*. Calcutta: West Bengal District Gazetteers.

Irrigation and Waterways Department n.d. *Sundarbaner Bhumikshoy o Nadibund Rakshanabekshane Sech o Jalapath Bibhagh* (The Irrigation and Waterways Department in Matters Concerning Land Erosion and Maintenance of Embankments in the Sundarbans). Calcutta: Government of West Bengal.

———. n.d.a. *Reconstruction and Rehabilitation Reports of Irrigation Departments of South 24 Parganas*. West Bengal: Government of West Bengal.

Ito, S. 2002. 'From Rice to Prawns: Economic Transformation and Agrarian Structure in Rural Bangladesh'. *The Journal of Peasant Studies* 29 (2): 47–70.

IUCN. 2009. 'Prawn Farms, the Supreme Court and IUCN: On-the-Ground Impediments to Conservation' *IUCN BMZ Project Case Studies*: 1–3.

Jackson, C. 1993a. 'Doing What Comes Naturally? Women and Environment in Development'. *World Development*, 21 (12): 1947–63.

———. 1993b. 'Women/Nature or Gender/History? A Critique of Ecofeminist "Development"'. *The Journal of Peasant Studies* 20(3): 389–419.

———. 1993c. 'Environmentalisms and Gender Interests in the Third World'. *Development and Change* 24 (4): 649–77.

———. 1998. 'Rescuing Gender From the Poverty Trap'. In *Feminist Visions of Development: Gender, Analysis and Policy*, edited by C. Jackson and R. Pearson. London: Routledge.

Jaladas, N. 2013. 'Fishermen, the "Forest Acts" and the Narratives of Eviction from Jambudwip Island'. In *NMML Occasional Paper Perspectives in Indian Development New Series 6*. New Delhi: Nehru Memorial Museum and Library Teen Murti.

Jalais, A. 2005. "Dwelling on Morichjhanpi: When Tigers Became 'Citizens' and Refugees 'Tiger food'". *Economic and Political Weekly*, 40(17): 1757–62.

———. 2010. *Forest of Tigers: People, Politics & Environment in the Sundarbans*. New Delhi: Routledge.

Jalil, A. F. M. A. 2000. *Sundarbaner Itihas* (A History of the Sundarbans). Calcutta: Naya Udyog.

Jeffery, R. and N. Sundar. 1999. 'Introduction'. In *A New Moral Economy for India's Forests?: Discourses of Community and Participation*, edited by R. Jeffery and N. Sundar. New Delhi, London: Sage Publications.

Johari, R. 2007. 'Of Paper Tigers and Invisible People: The Cultural Politics of Nature in Sariska'. In *Making Conservation Work: Securing Biodiversity in this New Century*, edited by G. Shahabuddin and M. Rangarajan. Ranikhet Cantt: Permanent Black.

Kabeer, N. 1994. *Reversed Realities: Gender Hierarchies in Development Thought*. London, New York: Verso.

Kanjilal, T. 2000. *Who Killed the Sundarbans?* Calcutta: Tagore Society for Rural Development.

Kohli, A. 1990. 'From Elite Activism to Democratic Consolidation: The Rise of Reform Communism in West Bengal'. In *Dominance and State Power in Modern India: Decline of Social Order*, vol. 2, edited by F. R. Frankel and M. S. A. Rao. Delhi: Oxford University Press.

———. 1997. 'From Breakdown to Order: West Bengal'. In *State and Politics in India*, edited by P. Chatterjee. Delhi: Oxford University Press.

Lahiri-Dutt, K. and G. Samanta. 2014. *People and Life on the Chars of South Asia: Dancing with the Rivers*. New Delhi: Foundation Books.

Lewis, D. J., R. Gregory, and G. D. Wood. 1993. 'Indigenising Extension: Farmers, Fish-Seed Traders and Poverty-Focused Aquaculture in Bangladesh'. *Development Policy Review*, 11: 185–94.

Mallick, R. 1993. *Development Policy of a Communist Government: West Bengal Since 1977*. Cambridge: Cambridge University Press.

———. 1999. 'Refugee Settlement in Forest Reserves: West Bengal Policy Reversal and the Marichjhapi Massacre'. *The Journal of Asian Studies*, 58 (1): 104–25.

Mies, M. and V. Shiva. 1993. *Ecofeminism*. London: Zed Books.

Ministry of Water Resources, Government of India. 2009. *Final Report of Task Force on Restoration of Sunderbans Embankments Damaged by the Cyclone 'AILA' in May 2009*. Delhi: Ministry of Water Resources.

Mohapatra, P. P. 1985. 'Coolies and Colliers: A Study of the Agrarian Context of Labour Migration from Chotanagpur, 1880–1920'. *Studies in History* 1 (2) n.s.: 247–303.

Mollah, A. R. 1993. *Budget Speech of the Minister-in-charge of Sundarban Affairs Branch, Development and Planning Department 1993–94*. Calcutta: Government of West Bengal.

———. 1994. *Budget Speech of the Minister-in-charge of Sundarban Affairs Department 1994–95*. Calcutta: Government of West Bengal.

———. 1998. *Budget Speech of the Minister-in-charge of Sundarban Affairs Department 1998–99*. Calcutta: Government of West Bengal.

Mondal, J. 2002. *Marichjhanpi: Noishabder Antaraley Ganahatya-r ek Kalo Itihas* (Marichjhanpi: Hidden Behind the Veiled Silence a Dark History of Genocide). Calcutta: People's Book Society.

Mondal, K. 1997. *Dakshin Chabbis Parganar Anchalik Itihaser Upakaran* (The Ingredients of the Local History of South 24 Parganas). Calcutta: Nabachalantika.

———. 1999. *Dakshin Chabbis Parganar Bismrita Adhyay* (A Forgotten Chapter in the History of South 24 Parganas). Calcutta: Nabachalantika.

Mondal, S. 1995. *British Rajotye Sundarban* (The Sundarbans During the British Rule). Calcutta: Punascha.

Montgomery, S. 1995. *Spell of the Tiger: The Man-Eaters of Sundarbans*. Boston, New York: Houghton Mifflin Company.

Mosse, D. 2003. *The Rule of Water: Statecraft, Ecology and Collective Action in South India*. Delhi: Oxford University Press.

Mukhopadhyay, A. 2009. 'On the Wrong Side of the Fence: Embankment, People and Social Justice in the Sundarbans'. In *Social Justice and Enlightenment: West Bengal*, vol. 1, edited by P. K. Bose and S. K. Das. New Delhi: Sage Publications.

———. 2010. 'Rethinking Community: In the Discipline and within Development Practices'. In *Sociology in India: Intellectual and Institutional Practices*, edited by M. Chaudhuri. New Delhi: Rawat Publications.

———. 2011. 'In Aila-struck Sundarbans'. *Economic and Political Weekly*, 46(40):21–24.

Nanda, K. 1998. *Budget Speech of the Minister-in-charge of Fisheries Department 1998–99*. Calcutta: Government of West Bengal.

————. 1999a. 'Sundarbans – A Unique Ecosystem: Now it is Threatened'. In *Sundarbans Mangal*, edited by D. N. Guha Bakshi, P. Sanyal, and K.R. Naskar. Calcutta: Naya Prokash.

————. 1999b. *Budget Speech of the Minister-in-charge of Fisheries Department 1999–2000*. Calcutta: Government of West Bengal.

————. 2004. *Budget Speech of the Minister-in-charge of Fisheries, Aquaculture, Aquatic Resource and Fishing Harbours Department 2004–2005*. Calcutta: Government of West Bengal.

————. 2005. *Budget Speech of the Minister-in-charge of Fisheries, Aquaculture, Aquatic Resource and Fishing Harbours Department 2005–2006*. Calcutta: Government of West Bengal.

————. 2009. *Budget Speech of the Minister-in-charge of Fisheries, Aquaculture, Aquatic Resource and Fishing Harbours Department 2009–2010*. Calcutta: Government of West Bengal.

Nandy, A. 1988. 'Science as the Reason of the State'. In *Science, Hegemony and Violence: A Requiem for Modernity*, edited by A. Nandy. Delhi: Oxford University Press.

————. 1990. 'The Politics of Secularism and the Recovery of Religious Tolerance'. In *Mirrors of Violence: Communities, Riots and Survivors in South Asia*, edited by V. Das. New Delhi: Oxford University Press.

Naskar, S. 2006. *Budget Speech of the Minister-in-charge of Irrigation and Waterways Department 2006–07*. Calcutta: Government of West Bengal.

————. 2007. *Budget Speech of the Minister-in-charge of Irrigation and Waterways Department 2007–08*. Calcutta: Government of West Bengal.

————. 2008. *Budget Speech of the Minister-in-charge of Irrigation and Waterways Department 2008–09*. Calcutta: Government of West Bengal.

————. 2009. *Aila Biddhasta Sundarban-e Durgato Manushder Parabarti Bhora Kotal Arthat 23.06.2009. o Asanna Barshar Prokop Theke Nirapatta-r Bishoye Mananiyo Bidhayak Sri Chittaranjan Mondal, Jane Alam Mian, Dhananjoy Modak o Narmada Chandra Roy er Utthapito Drishti Akarshani Prastabe-r Prekshite Sech o Jalapath Bibhager Mantri Sri Subhas Naskar er Bibriti. Prastab ti 16.06.2009 Tarikhe Utthapito Hoi* (The Irrigation and Waterways Department Minister Sri Subhas Naskar's Account in Response to the Attention Drawn by MLAs such as Sri Chittaranjan Mondal, Jane Alam Mian, Dhananjoy Modak and Narmada Chandra Roy Towards How to Provide Safety for the Aila-Stricken People of the Sundarbans Against Rise in Water During the Next High Tide on Full Moon i.e., 23.06.2009 and the Ensuing Monsoon. This Proposal Was Tabled on 16.06.2009). The West Bengal Legislative Assembly Session, Calcutta, June 16.

————. 2010. *Budget Speech of the Minister-in-charge of Irrigation and Waterways Department 2010–11.* Calcutta: Government of West Bengal.

Nayek, A. 2005. 'Nadibund O Sundarbans' (Embankments and the Sundarbans). *Badweep Barta,* Canning, October 16–31.

Nelson, N. and S. Wright, eds. 1995. *Power and Participatory Development: Theory and Practice.* London: Intermediate Technology Publications.

Nicholas, R. W. 1963. 'Ecology and Village Structure in Deltaic West Bengal'. *Economic Weekly.* Special no. July: 1185–96.

Nugent, D. 1994. 'Building the State, Making the Nation: The Bases and Limits of State Centralisation in "Modern" Peru'. *American Anthropologist* 96 (2): 333–69.

Oldenburg, V. T. 1991. 'Lifestyle as Resistance: The Case of the Courtesans of Lucknow'. In *Contesting Power: Resistance and Everyday Social Relations in South Asia,* edited by D. Haynes and G. Prakash. Berkeley Los Angeles: University of California Press.

Oliver-Smith, A. 1977a. 'Traditional Agriculture, Central Places and Postdisaster Urban Relocation in Peru'. *American Ethnologist* 4 (1): 102–16.

————. 1977b. 'Disaster, Rehabilitation and Social Change in Yungay, Peru'. *Human Organization* 36 (1): 5–13.

————. 1991. 'Successes and Failures in Post-Disaster Resettlement'. *Disasters* 15 (1): 12–24.

————. 1996. 'Anthropological Research on Hazards and Disasters'. *Annual Review of Anthropology* 25: 303–28.

O'Malley, L. S. S. 1913. *The Census of India 1911: Bengal, Bihar, Orissa and Sikkim,* vol. 5, Part I. Calcutta: Bengal Secretariat Book Depot

————. 1998 [1914]. *Bengal District Gazetteers: 24 Parganas.* Calcutta: West Bengal District Gazetteers.

Ortner, S. B. 1974. 'Is Female to Male as Nature Is to Culture?' In *Woman, Culture, and Society,* edited by M. Z. Rosaldo and L. Lamphere. California: Stanford University Press.

Ortner, S. B. and Whitehead, H. 1981. 'Introduction: Accounting for Sexual Meanings'. In *Sexual Meanings: The Cultural Construction of Sexuality,* edited by S. B. Ortner and H. Whitehead. Cambridge: Cambridge University Press.

Pal, M. 2009. *Marichjhanpi: Chhinna Desh, Chhinna Itihas* (Marichjhanpi: Fragmented Homeland, Fragmented History). Calcutta: Gangchil.

Pande, M. 1998. *'Foreword', Annual Administrative Reports 1996–97, 1997–98, 1998–99,* Department of Sundarban Affairs. Calcutta: Government of West Bengal.

Pargiter, F. E. 1934. *A Revenue History of the Sundarbans: From 1765 to 1870.* Bengal: Bengal Government Press.

Paul, S. 1987. *Community Participation in Development Projects*. World Bank Discussion Paper no. 6. Washington, DC: World Bank.

Prakash, G. 1992. 'Science "Gone Native" in Colonial India'. *Representations* 40: 153–77.

Rai, N. D. 2007. 'The Ecology of Income: Can We Have Both Fruit and Forest?' In *Making Conservation Work: Securing Biodiversity in this New Century*, edited by G. Shahabuddin and M. Rangarajan. Ranikhet Cantt: Permanent Black.

Randeria, S. 2010. 'Opting for Statelessness' (Review of James Scott 'The Art of Not Being Governed: An Anarchist History of Upland Southeast Asia'). *European Journal of Sociology* 51(3): 464–9. doi:10.1017/S0003975610000329. Accessed November 2011.

Rangarajan, M. and G. Shahabuddin. 2007. 'Introduction'. In *Making Conservation Work: Securing Biodiversity in this New Century*, edited by G. Shahabuddin and M. Rangarajan. Ranikhet Cantt: Permanent Black.

Rangarajan, M. and K. Sivaramakrishnan. 2012. 'Introduction'. In *India's Environmental History: Colonialism, Modernity and the Nation* vol. 2, edited by M. Rangarajan and K. Sivaramakrishnan. Ranikhet Cantt: Permanent Black.

Richards, J. F. and E. P. Flint. 1987. 'Long Term Transformations in the Sundarbans Wetlands Forests of Bengal'. Paper presented in *The Commons in South Asia: Societal Pressures and Environmental Integrity in the Sundarbans*, the workshop held at Smithsonian Institution, Washington, DC. November 20–21. www.smartoffice.com/Tiger/Richards.html. Accessed September 2002.

Rodgers, A., D. Hartley, and S. Bashir. 2005. 'Community Approaches to Conservation: Some Comparisons from Africa and India'. In *Battles Over Nature: Science and the Politics of Conservation*, edited by V. Saberwal and M. Rangarajan. Delhi: Permanent Black.

Rogers, J. D. 1991. 'Cultural and Social Resistance: Gambling in Colonial Sri Lanka'. In *Contesting Power: Resistance and Everyday Social Relations in South Asia*, edited by D. Haynes and G. Prakash. Berkeley, Los Angeles: University of California Press.

Ruud, A. E. 2001. 'Talking Dirty About Politics: A View From a Bengali Village'. In *The Everyday State and Society in Modern India*, edited by C. J. Fuller and V. Beneï. London: Hurst and Company.

Samaddar, R. 1999. *The Marginal Nation: Transborder Migration from Bangladesh to West Bengal*. New Delhi: Sage.

Samaddar, R. P. 2000. 'Introduction', *Annual Administrative Reports 1996–97, 1997–98, 1998–99*. Department of Sundarban Affairs. Calcutta: Government of West Bengal.

Sarkar, S. K. and A. K. Bhattacharya. 2003. 'Conservation of Bio-diversity of the Coastal Resources of the Sundarbans, Northeast India: An Integrated Approach through Environmental Education'. *Marine Pollution Bulletin* 47: 260–64.

Scott, J. C. 1985. *Weapons of the Weak: Everyday forms of Peasant Resistance.* New Haven and London: Yale University Press.

———. 1990. *Domination and the Arts of Resistance: Hidden Transcripts.* New Haven and London: Yale University Press.

———. 2000. 'Foreword'. In *Agrarian Environments: Resources, Representations, and Rule in India,* edited by A. Agrawal and K. Sivaramakrishnan. Durham, London: Duke University Press.

———. 2009. *The Art of Not Being Governed: An Anarchist History of Upland Southeast Asia.* New Delhi: Orient Blackswan.

Sharma, A. and A. Kabra. 2007. 'Displacement as a Conservation Tool: Lessons from the Kuno Wildlife Sanctuary, Madhya Pradesh'. In *Making Conservation Work: Securing Biodiversity in this New Century,* edited by G. Shahabuddin and M. Rangarajan. Ranikhet Cantt: Permanent Black.

Sharma, C. 1998. 'Aquaculture in India: The Supreme Court Verdict'. http://base.d-p-h.info/en/fiches/premierdph/fiche-premierdph-4040.html Accessed August 2014.

Shiva, V. 1986. 'Ecology Movements in India'. *Alternatives* 11(2): 255–73.

———. 1988. *Staying Alive: Women, Ecology and Survival in India.* London: Zed Press.

Shore, C. and S. Wright. 1997. 'Policy: A New Field of Anthropology'. In *Anthropology of Policy: Critical Perspectives on Governance and Power,* edited by C. Shore and S. Wright. London: Routledge.

Simpson, E. and S. Corbridge. 2006. 'The Geography of Things That May Become Memories: The 2001 Earthquake in Kachchh Gujarat and the Politics of Rehabilitation in Prememorial Era'. *Annals of the Association of the American Geographers* 96 (3): 556–85.

Simpson, E. 2007. 'State of Play Six Years after Gujarat Earthquake'. *Economic and Political Weekly* 42 (11): 932–37.

Singh, K. S. 1966. *The Dust Storm and the Hanging Mist: A Study of Birsa Munda and His Movement in Chotanagpur 1874–1901.* Calcutta: FIRMA K.L. Mukhopadhyay.

———. 1983. *Birsa Munda and His Movement 1870–1901: A Study of a Millenarian Movement in Chotonagpur.* London, Calcutta: Oxford University Press.

Singh, R. K. 1988. 'How Can We Help the Poorest of the Poor? The Impact of Local Elites in Some NGO Community Development Programmes in

Nepal and Bangladesh'. In *Putting People First: Voluntary Organisations and Third World Development*, edited by R. Poulton and M. Harris. London: Macmillan Publishers.

Sinha, S, S. Gururani, and B. Greenberg. 1997. 'The "New Traditionalist" Discourse of Indian Environmentalism'. *The Journal of Peasant Studies* 24 (3): 65–99.

Sivaramakrishnan, K. 1997. 'A Limited Forest Conservancy in Southwest Bengal, 1864–1912'. *The Journal of Asian Studies* 56 (1): 75–112.

———. 1999. *Modern Forests: Statemaking and Environmental Change in Colonial Eastern India*. New York, New Delhi: Oxford University Press.

Skaria, A. 1997. 'Shades of Wildness: Tribe, Caste and Gender in Western India'. *The Journal of Asian Studies* 56 (3): 726–45.

Sundarban Affairs Department. 1994. *Annual Administration Reports 1981–82 to 1991–92*, Department of Sundarban Affairs. Calcutta: Government of West Bengal.

———. 2000. *Annual Administrative Reports 1996–97, 1997–98, 1998–99*, Department of Sundarban Affairs. Calcutta: Government of West Bengal.

———. 2005. *Annual Administrative Reports 2004–5*, Department of Sundarban Affairs. Calcutta: Government of West Bengal.

Sundarban Development Board. 1979. *A Project Report Prepared by the Sundarban Development Board and the Working Group of Government of West Bengal in collaboration with Central Project Preparation and Monitoring Cell Ministry of Agriculture and Irrigation, Government of India*. Calcutta: Sundarban Development Board.

———. 1979a. *Sundarban Region – Development Activities Carried out by the Sundarban Development Board: A Profile*. Calcutta: Sundarban Development Board.

Tarlo, E. 2001. 'Paper Truths: The Emergency and Slum Clearance Through Forgotten Files'. In *The Everyday State and Society in Modern India*, edited by C. J. Fuller and V. Bénéï. London: Hurst and Company.

Taussig, M. 1984. 'Culture of Terror – Space of Death. Roger Casement's Putumayo Report and the Explanation of Torture'. *Comparative Studies in Society and History* 26 (3): 467–97.

———. 1991. *Shamanism, Colonialism and the Wild Man: A Study in Terror and Healing*. Chicago: The University of Chicago Press.

The Statesman. 2002. 'Indo-Bangla effort to save the Sunderbans'. *The Statesman*, Kolkata, April 4.

The Telegraph. 2009a. 'Rs 1000 cr Aila aid to state govt'. *The Telegraph*, Kolkata, July 7.

————. 2009b. 'Delhi gives Rs 478 cr for Aila repairs'. *The Telegraph*, Kolkata, July 8.

————. 2009c. 'Acquisition for Aila dykes'. *The Telegraph*, Kolkata. November 5.

————. 2013. 'Politics of Aila, Caked in Mud and Apathy – House and River Barrier Grouse'. *The Telegraph* (Bengal Supplement). July 19.

The Times of India. 2009. 'Politics over Aila funds'. *The Times of India*, Kolkata, October 30.

The Week. 1989. 'A Storm and a Messiah: The Heroic Struggle of a Social Worker and a people'. *The Week*, Kolkata, 5-11 February.

United Nations. 1951. *Measures for the Economic Development of Under-developed Countries*, UN Department of Economic Affairs. New York: United Nations.

————. 1993. *Report of the United Nations Conference on Environment and Development at Rio de Janeiro, June 3–14, 1992*, vol. 1. New York: United Nations.

Vasavada, S, A. Mishra, and C. Bates. 1999. 'How Many Committees Do I Belong To?' In *A New Moral Economy for India's Forests? Discourses of Community and Participation*, edited by R. Jeffery and N. Sundar. New Delhi, London: Sage Publications.

Vedeld, T. 2001. *Participation in Project Preparation: Lessons from World Bank-Assisted Projects in India*, Discussion Paper no. 423. Washington, DC: World Bank.

Vidal, J. 2003. 'It's Impossible to Protect the Forests for Much Longer'. *The Gurdian*, July 31.

Visvanathan, S. 1988. 'On the Annals of the Laboratory State'. In *Science, Hegemony and Violence: A Requiem for Modernity*, edited by A. Nandy. Delhi: Oxford University Press.

————. 1991. 'Mrs Brundtland's Disenchanted Cosmos'. *Alternatives* 16 (3): 377–84.

Wade, R. 1982. 'The System of Administrative and Political Corruption: Canal Irrigation in South India'. *Journal of Development Studies* 18 (3): 287–328.

West Bengal. 2000. *Paschimbanga: Jela Dakshin Chabbis Pargana* (West Bengal: The District of South 24 Parganas), vol. 33, no. 32–36, February-March 2000. Calcutta: Department of Information and Culture, West Bengal Government.

Whitcombe, E. 1995. 'The Environmental Costs of Irrigation in British India: Waterlogging, Salinity and Malaria'. In *Nature, Culture, Imperialism: Essays on the Environmental History of South Asia*, edited by D. Arnold and R. Guha. New Delhi: Oxford University Press.

Wilson, E. O. 1995. 'Loss of Biodiversity' in *Meeting the Challenges of Population, Environment and Resources: The Costs of Inaction*, A Report of the Senior Scientists' Panel, Third World Bank Conference on *Environmentally Sustainable Development* (ESD Proceeding Series no. 14), October 4–9. Washington, DC: World Bank.

World Bank. 1994. *Enhancing Women's Participation in Economic Development: A World Bank Policy Paper*. Washington, DC: World Bank.

———. 1996a. *The World Bank Participation Source-Book*. Environmentally Sustainable Development (ESD). Washington, DC: World Bank.

———. 1996b. *The World Bank's Partnership with Nongovernmental Organizations*. Participation and NGO Group, Poverty and Social Policy Department (PSP). Washington, DC: World Bank.

———. 2003. 'Sustainable Development in a Dynamic World: Transforming Institutions, Growth, and Quality of Life'. *World Development Report*. Washington, DC and New York: World Bank and Oxford University Press.

Government of India census websites

http://www.censusindia.gov.in/pca/SearchDetails.aspx?Id=353079
http://www.censusindia.gov.in/pca/SearchDetails.aspx?Id=354440
http://www.censusindia.gov.in/pca/SearchDetails.aspx?Id=354647
http://www.censusindia.gov.in/pca/SearchDetails.aspx?Id=354523
http://www.censusindia.gov.in/pca/SearchDetails.aspx?Id=354610
http://www.censusindia.gov.in/pca/SearchDetails.aspx?Id=342254
http://www.censusindia.gov.in/pca/SearchDetails.aspx?Id=352326
http://www.censusindia.gov.in/pca/SearchDetails.aspx?Id=353125
http://www.censusindia.gov.in/pca/SearchDetails.aspx?Id=354565
http://www.censusindia.gov.in/pca/SearchDetails.aspx?Id=350503
http://www.censusindia.gov.in/pca/SearchDetails.aspx?Id=353025
http://www.censusindia.gov.in/pca/SearchDetails.aspx?Id=350401
http://www.censusindia.gov.in/pca/SearchDetails.aspx?Id=354493
http://www.censusindia.gov.in/pca/SearchDetails.aspx?Id=341715
http://www.censusindia.gov.in/pca/SearchDetails.aspx?Id=341577
http://www.censusindia.gov.in/pca/SearchDetails.aspx?Id=341610
http://www.censusindia.gov.in/pca/SearchDetails.aspx?Id=341637
http://www.censusindia.gov.in/pca/SearchDetails.aspx?Id=340107
http://www.censusindia.gov.in/pca/SearchDetails.aspx?Id=341497
Websites Accessed between 31 May to 20 June 2014

Index